TELECOM
MANAGEMENT
FOR LARGE ORGANIZATIONS

A Practical Guide

Luiz Augusto de Carvalho

TELECOM MANAGEMENT FOR LARGE ORGANIZATIONS
A PRACTICAL GUIDE

iUniverse books may be ordered through booksellers or by contacting:

iUniverse
1663 Liberty Drive
Bloomington, IN 47403
www.iuniverse.com
1-800-Authors (1-800-288-4677)

ISBN: 978-1-5320-3280-6 (sc)
ISBN: 978-1-5320-3279-0 (e)

Library of Congress Control Number: 2017915483

Print information available on the last page.

iUniverse rev. date: 10/14/2017

Contents

Preface

As a telecommunication Engineer and Business administrator specialized in wide area network design working continuously in this field since 1989 I perceive the need for a practical guide addressing the most common issues associated with managing the telecom area in a large organization.

This book is aimed at all professionals linked with the IT/Telecom management within large organizations and I hope to have produced a good guideline to help you to navigate through these issues.

The book is divided into twenty four chapters, each of which deals with a specific aspect of managing large telecommunications infrastructures; it falls into five sections: The telecom manager, macro issues, telecom consulting, bills and billing processing, and the future.

The first two chapters are concerned with the role of telecom managers in large organizations. Chapters 1 discusses how the telecom manager tpically works and try to speculate about the reasons that makes it more or less effcient and discusses strategies for properly alocating the managing effort, which in our view is a key aspect.

1. The Telecom Manager
2. The technical capability

The next section is concerned with the main issues associated with managing the telecom structure in a large organization. Chapter 3 discusses how to evaluate properly the correlation control effort vs. control benefits. Chapter 4 and 5 deal with the issue of how to define the priorities properly avoiding undue influences from the vendors. Chapter 6 and 7 discuss the specific aspects that make negotiating telecom contracts different than the typical IT contract and how to negotiate them. Chapter 8 deals with the pros and cons of outsourcing the telecom management in a large organization. Chapter 9 and 10 discuss managing the inventory of assets and services, which in our view is the basis over which all control is developed. Chapter 11 and 12 deal with the process of adding, removing, and changing the assets and services contracted and why it is usually a source of inefficiency.

3. The perfect the worst enemy of the good
4. Hijacked agenda
5. How to define priorities
6. Why negotiating telecom contracts is different
7. Negotiating with the Service Providers
8. The issue of business process outsourcing
9. Inventory control – why it is so difficult
10. Asset and Service inventory Management
11. The interaction with the providers a big source of inefficiency
12. Service Ordering & Changing control

The third section discusses what is more commonly known as telecom consulting and the aim is to discuss issues related to contracting and providing telecom consulting services. Chapter 13 discusses how to make a quick assessment, chapter 14 focuses on how to deal with external consultants, and chapter 15 discusses the need for external consultants.

13. How to make a quick assessment
14. How to deal with external consultants
15. Why use external consultants

The fourth section deals with issues linked with telecom expense management (TEM). Chapter 16 discusses where to fit the telecom bill processing area within the organization. Chapter 17 deals with the relative importance of auditng bills. Chapter 18 discusses why the deep roots of the billing errors very often are not treated. Chapter 19 deals with the issue of automatizing the bill processing and chapter 20 describes the main processes associated with managing the bills.

16. Where to fit the telecom bill processing area within the organization
17. Auditing bills unbalanced importance
18. Billing errors why they don't fix the problem
19. Automatizing the bill processing
20. Invoice processing – How to manage the bills

The final section discusses the scenario to where the telecom infrastructure is moving to try to foresee how the telecom role will look like in the near and medium future.

21. Telecom management in an evolving scenario

This book is not meant to be read linearly like a story book; it is more like a manual, where you can jump directly to the chapter dealing with a particular issue. Of course, a linear reading is possible and even desirable, given the fact that earlier chapters support or complement later ones. I hope you enjoy reading it; I also hope it will become a useful instrument for you. This area of expertise lacks a broad source of literature, and the lessons learned by the professionals in the field are rarely documented or shared. I made an honest effort to document and systematize this practical knowledge.

1 The Telecom Manager – How He/She Works

One thing which always impressed me is how difficult the job of a telecom manager is (I know it firsthand having worked in this position for almost five years in a large bank with 400 branches). They usually are a very busy bunch of people. In my long professional life I had chance to observe several of them at work. From these observations I tried to understand what the main drives of this particular position are.

Here it is worth mentioning that the term "telecom manager" here, refers to the people in charge of the telecom infrastructure in large organizations, usually with monthly expenditures with telecom above USD 500.000,00.

In terms of position within the organization, the more usual scenario is one where you have the IT manager, under who there is an infrastructure manager under who there is the telecom manager or telecom coordinator. Therefore the usual scenario is having the telecom manager/coordinator one or two levels below the IT manager. In organizations with a very heavy dependency from telecom we may have scenarios where the telecom manager responds directly to the IT manager.

My intention is to identify what are the features which make a telecom manager more efficient developing his work. It is interesting to clarify what I mean by "efficient". It may sounds obvious but sometimes isn´t. Usually, a telecom manager who keeps the structure running without problems and manage to implement the new services without major delays is considered efficient. In our view however it is just half of the history. The ideal telecom manager has to be able to not only doing these two things but do them keeping an eye in the costs and plan ahead considering the future needs and the new technologies and services available.

The first thing to be noticed is the fact that these people usually have four main lines of activities competing for their attention:

1) The issues linked with keeping the structure operating
2) The issues linked with new resources/services being quoted and implemented
3) The issues linked with managing their teams
4) The issues linked with controlling the structure and providing information for their senior management and planning for the future.

We are going to discuss the difficulties associated with adjusting the attention and time of the telecom manager and his team to the real needs of the organization.

The first and biggest dilemma is how to allocate the time of the technical teams between operational problems and new projects. Some organizations are so big or so dynamic that they may afford to have two different teams but usually you have only one telecom team responsible for everything.

The second typical dilemma is how to define which projects should be prioritized. It may sound a minor issue but there are several forces acting to make this very difficult.

New projects usually can be divided between expansion (Ex: New sites) and implementation of new services. New sites usually imply in deadlines defined by the business what makes, as time goes and the deadlines approach, these activities become naturally priorities.

New services in their turn are influenced by a wide range of factors not all of them under the control of the telecom manager:

1) Dependency from other areas within and without the organization

2) Prioritization defined by the IT manager or the infrastructure manager

3) Level of effort demanded (internal or external)

Subsequently there are the administrative tasks linked with managing the teams. These may sound a bit mundane but they are tasks which sometimes demand a lot of time. Things such as working schedule, controlling the overtime and expenses all demand attention and cannot be avoided.

Adding to these we have the episodic demands of the high management, these also usually demand a lot of time and cannot be avoided.

1.1 How to properly allocate the time?

The key to properly allocate time is proper planning conjugated with an adequate cost control. It is not to say that we can plan everything, however, having some sort of strategy and keeping an eye in the structure costs helps to guide you through the day by day demands.

Once a year is important to make a telecom plan, the ideal is to do that together with budget planning for the next year. Summarizing, when planning the budget you have to plan all initiatives you expect to take regards telecom for the next year and their economic implications. This plan is going to be a guideline for your activities and therefore will guide you through the year setting priorities and goals.

This plan must include the expansions (Assuming the best available information provided by the business) and all actions foresaw for the telecom structure.

The definition of these actions must be based on the current scenario evaluation identifying all possible actions in order to achieve quality improvement and cost reductions.

To each action you have to have a timeframe (indicating if the action has a short, medium or long term impact) and of course the economic impact of the action.

Therefore the plan has to have the actions associated to expansion and new projects. This plan should be ordered by level of difficult vs. economic benefits. This ordering is important because sets the priority of each action. Of course this prioritization is not absolute in a sense that you can always change it along the way, however having a basic prioritization helps a lot and avoid getting lost in the middle of the daily activities of competing demands.

Usually these lines of activities demand 40%, 40%, 10% and 10% of their time respectively. Of course it varies along the time (During the rollout of a new network the percentile of the time dedicated to new resources may go to 80% for instance).

2 The Technical Capability

Here it is worth mentioning that technical knowledge isn't the most important quality in a good telecom manager. That doesn't mean that a good telecom manager doesn't need to have a good understanding of the issues and technologies involved in his work, but it is just a part of the whole set of skills demanded for the position. It may sounds a strange statement and counter intuitive given the fact that every job description for these positions are full of technical requirements. In my view having a good understanding of the technical issues is important but more important than that is being able to put these issues in the context of the organization as whole (Contracts, prices, demands for services, who uses what, what happens if something fails and so on) and be able to manage the teams, the providers and the time properly.

This issue becomes more acute because very often the technicians and engineers who perform better are promoted to managerial positions where the skills and features that made them good technicians may work against them being good managers. If you know how to do a thing yourself you may fell tempted to solve the problem yourself instead training somebody to do so or paying for it. Very often we see telecom managers lost between their former technical tasks and the managerial functions. That usually generates a lot of frustration.

A good telecom manager has a good understanding of the technology deployed but uses logic as the preferred tool, putting together the relations of cause and effect of each phenomenon. That attitude usually points to the location of the problems (even when the person is not very technical).

The same ability to use logic helps when thinking about the structure and planning for the future. In other words a person to perform well in this position has to be able to think the solution as whole not necessarily knowing how to work with each piece of equipment/software.

Adding to this view we have to keep in mind that even when the telecom area is managed with high dree of efficiency if it is not operating in line with the objectives and strategies of the organization as whole that doesn't work well. The telecom manager has to understand that he/She is just a small part in a much big machine. Too technical people tend to lose this perspective.

In summary, a good telecom manager has not only good technical knowledge but also ability to manage and motivate people, be able to use logic, have a good judgment and be aligned with the objectives of the organization as whole. Of course this view applies not specifically to the telecom manager, probably it holds true for every management position.

3 The Perfect as the Worst Enemy of the Good

An impressive thing when dealing with large Consulting companies and the techniques to make assessments and ITIL best practices is how difficult becomes to put what they give you in the right perspective. It is not to say that these things don't have value, however, when you look the problems most companies have, things usually are too basic.

For example you don't need a Gartner consultancy to tell you that you have to have an inventory of your resources to be able to match it with the bills you are receiving to check if the resources you are paying belong to your organization. Please don't get it wrong it doesn't mean that what they say is worthless, the problem with these issues is the fact that much of what they give you is just common sense, which anybody with the minimal good sense will be able to perceive and fix.

In addition of that, when you bring in these methodologies and consultancies sometimes the whole thing becomes too complicate. It happens because you are presented with a whole set of adjustments to bring you from 0% to 100% of compliance, (consultants are going to say that the thing is scalable), that transforms what could be a small scale adjustment in a big and complex project. Sometimes this process makes the problem too complex and too expensive to be solved easily; most times a piece meal approach would yield much better and fast results.

Very often a small scale initiative which could be addressed by the telecom team internally with a minimal effort and cost, becomes part of a bigger initiative encompassing many other issues (And parts of the organization).

The budget bracket increases (escaping from the telecom manager) and the whole thing become much more complex and difficult increasing the chances that the problem may not be solved at all. (true it may solve other issues if succeed). Maybe this is an unsolvable paradox. However, in practice, we see more problems not being solved because the issue was rendered too complex than when becoming complex all problems are solved (SAP implementations are usually a typical example where this issue occurs).

This statement sounds a bit against the tide, but most times a small scale implementation works better and the right thing to do is to increase the level of sophistication of the control gradually. Most times small things make a huge difference. The right approach is to go in layers, doing the basic first and as time goes try to balance the act between control effort and gains.

When balancing the act between effort to control and benefits two things come to mind:

1) Granularity of the control
2) Complexity of processes

It demands good judgment from the people involved. Questions such as:

3) Makes sense to implement a billing system when we don't even have an inventory of trunks?
4) Makes sense to do billing auditing if we don't even have telco contracts with standardized tariffs?

Questions like that have to be made and answered. It sounds a bit obvious but you have to remember that these kinds of decision are taken by the senior manager under the pressure from the consultancies, service providers and hardware providers. Each one of them tends to inflate the importance of its own part. That reminds me a phrase:

"Just because you have a good hammer it doesn't mean that every problem is a nail".

Here it is important to be able to put things in perspective ordering the activities by the effort demanded and the benefits yielded.

An illustrative example is a case developed in Brazil where a large consulting company made an assessment about the TEM practices in a large enterprise and concluded that there was a lack of processes and policies regards mobile devices. This diagnosis was passed to the senior manager and it was set as a priority to the telecom area. When presented with this fact the telecom manager was able to demonstrate that although the facts mapped by the consulting company were true, mobile costs respond for just USD 60.000 monthly against a total expenditure with telecom of USD 1.500.000 (4%).

How much time and effort from the telecom team was worth investing to improve the control over these 4%? How much would be achievable in terms of cost reduction?

This kind of situation happens all the time, with the difference that frequently the people involved cannot put the problem in the right perspective. The people involved very often don't have a clear view of the relative weight of each issue in the overall expenditures. Nor the consultants, nor the telecom team (Typically when cost management are not part of their core responsibilities).

If we could summarize: Try to make the basics right and only when you archived this basic level try to move forward. The perfect usually is the worst enemy of the good.

For example: When managing contracts, never try to implement complex controls before you are able to manage the basics well. You must be able to control the cycles of the contracts (following when each contract ends) and the resources they cover and their costs. Once you have these basic aspects well controlled, you can start controlling things such as historical monthly costs and reimbursements due to noncompliance with the SLAs.

4 Hijacked Agenda

In jurisdictions where national (typically semi-government controlled) monopoly Telcos are the predominant telecommunications providers with low levels of competition, you have a situation where providers have a tendency to hijack the telecom manager agenda. In this item I discuss this phenomenon.

The telecom manager, in general, has a high level of dependency and interaction with the service providers. It is a symbiotic relationship whose day by day sometimes tends to make the views of the providers prevail over the views of the organization. (and its needs). This problem tends to be correlated with the availability of providers operating in a given market and with the ability of the telecom manager to understand the organizations needs. In other words less providers and less understanding makes the problem more acute.

The characteristics of this relationship sometimes masquerade the basic fact that there is a fundamental conflict of interest between the telecom manager and the providers. It is never too much to repeat:

"The objective of the organization is to achieve as much services as possible with the highest possible quality for the smaller possible cost. The objective of the provider is to provide the services with the least effort as possible for the maximum possible value."

The contracts reflect some sort of middle ground between these two extremes.

Although the previous statement sounds a bit obvious, the interdependence of the relationship very often leds to a scenario where the providers start inducing the direction and the rhythmus of the telecom projects within the organization. Very often It happens when you have a telecom manager unsure of his role conjugate with few and good telco representatives.

The telco representative becomes a friend to relay upon in face of the vicissitudes of the job. Who is your friend to call when an installation is not going as fast as it should?, Who is the person who comes with good ideas when you have none?

Very often this fundamental conflict of interest in this relationship is hidden under a speech of "partnership" which often opens the door for much more subtle forms of influence.

A good example of what that means is the issue of "being in the technological state of art". It reminds a sales statement made by a sales representative for a large organization about the benefits of migrating the current frame relay network for a brand new MPLS network. "You are going to have the latest technology paying very little more than what you pay today", when a spoiler made a simple question "Well, but what exactly we are going to gain from having this new technology?", the answer was "You are going to be updated with what is the latest and best technology" – The answer was hyperbolic and seems to imply that it is obvious that a new technology is better and worthy paying more for. The telecom manager consented…

This strange conversation depicts the problem. Without entering in the discussion about the specific benefits made available by the MPLS technology, it is not obvious (as the sales person implies and the telecom manager concurs) that a new network would bring any immediate benefit for the organization, even less obvious that it would be worthy paying more for it.

In the bottom of this discussion is the fact that the provider had implemented a new MPLS network and needed to migrate its current clients to this network. The benefit was basically for the telco but was sold as if it was a need of the client. "Don't be left behind with an old technology".

The intention here is not so much to discuss the specifics of this case, but use it as an example how the providers influence the organization decision making process. The telecom manager has to be aware of the phenomenon to avoid be carried away by it. The hardware and services providers tend to create a kind of consensus about what has to be done and convince you that. In this case the consensus is "The new technology is always better, and you should be always in the state of art".

Of course the phenomenon is not exclusive to the telecom environment (It remembers me the real need we have to update our windows). But the fact remains, even if you realize that at some point you are going to have to migrate to new technologies (even if only because the provider stops to support the technology you use) you should do it in your own terms and not let the provider dictate the rhythmus.

In this case the provider tried to create a need. Of course this strategy holds true for almost all types of sales and it isn't specific for the telecom environment. The basic idea is to create the need. The problem happens when the sway of this kind of idea takes the attention of the telecom manager away from the real needs which may not be linked with having the latest technology. Of course there is no good wind if you don't know to where you should go, and if you don't know where to go, there will be always a provider to give you some idea.

But the thing is not only about bring you to develop needs that you may not have it is also about bringing up in your priority list items which otherwise would be very low.

Example of this kind of thing can be seen very often when dealing with the issue of billing systems and auditing bills.

For example, although having a good inventory of resources and consolidate contracts rates much higher than billing and auditing as needs in a large telecom area (in fact being prerequisite to them) the providers of such solutions very often tend to emphasize the need for auditing and billing (Emphasizing short term savings) even when they themselves know the organization doesn't have even the basics.

Therefore it is very important to understand the needs of your organization, the key point is the fact that the telecom manager has to be close to the business needs and has to have a very good economical view of the whole process. With this understand becomes easier to see the ROI of each initiative and chart the right course to be followed. With this plan ready verify which solution and provider fits better.

In other words, everything is linked with understanding the needs of your users and the business of your organization, analyzing your traffic and of course understanding the characteristics of the services and technologies available.

Last but not the least, It is very important to have other sources of information besides your own providers. Big users groups, ITIL discussion groups and association such as TEMIA are good examples.

That said we come back to the issue of planning. As I mentioned in previous items I strongly recommend the preparation of an annual telecom plan. This plan sets the priorities straight and gives you time to analyze what the needs actually are and what is just hype.

5 How to Define the Priorities

Initially it is important to distinguish between Telcos (the network is part of their core business) and other operations where the network is part of the infrastructure that makes things work. Here we are talking about organizations where the network is part of the infrastructure, therefore a mean not an end in itself.

Very often infrastructure is seen as commodity (Although sometimes an expensive one), It happens even when IT is seen as strategic. Strategic in this sense usually applies only to the applications not so much to the underlying infrastructure.

In this kind of organization the definition of priorities of the projects is usually defined in the higher level within an IT department and it is usually called portfolio management. This process looks at all the projects and IT plans and negotiate project priorities with the business, not strictly networking. However network projects may disrupt or influence other IT projects so the whole portfolio of projects are looked at as a whole and then resourced according to major impact to business.

Deciding how to prioritize and separate the high priority from lower priority projects can be difficult. In our view these kinds of decisions demands a structured and objective approach that helps to achieve a balancing between the needs of the business and the resources available to implement them.

Infrastructure changes, in general, requires testing and validation often involving many non infra professionals even within IT. In addition of that money always talks and capex and opex allocation is very important factors in the whole decision making exercise, often cost savings are used as the basis for infra projects justification, and even a way to justify other IT project not related to infrastructure, a matter of re-allocating financial resources.

Using a prioritization matrix is a proven technique for making tough decisions in an objective way. A prioritization matrix is a simple tool that provides a way to sort a diverse set of items into an order of importance. It also identifies their relative importance by deriving a numerical value for the priority of each item.

The matrix provides a means for ranking projects based on criteria that are determined to be important, this is exactly the kind of thing which makes or breaks a good annual telecom plan.

Through this technique the telecom area can see clearly which projects are the most important to focus on first, and stablish a ranking of the others.

In addition of that the Prioritization matrix establishes a platform for conversations about what is important within the IT area and between the IT area and the Business.

The first thing is to determine its criteria and weights those criteria based on savings achievable, organizational goals, available resources, and so on. Projects are then scored and prioritized based on the defined criteria. Once projects are prioritized and those priorities are reviewed and discussed, the telecom area can evaluate the results to determine, funding and resource allocation for the higher priority projects (in fact this is the annual telecom plan). A final step involves assessing how and when (or if) to fund the lower priority projects in the future if/when more resources become available.

Creating and using a prioritization matrix involves five simple steps:

1. Determine your criteria and rating scale. There are two components involved in rating the projects on your "to do" list: criteria for assessing importance, and a rating scale.

The first step is to determine the factors you will use to assess the importance of each project. Choose factors that will clearly differentiate important from unimportant projects – these are your criteria. A group of 4-8 criteria is typical. Example criteria might include whether or not the project is mandatory (Ex: The operation stops without it), the value it brings to the business, the savings it generates for the organization etc.

Then, for each of your criteria, establish a rating scale to use in assessing how well a particular project satisfies that criteria. To ensure consistent use of the rating scale, provide some details to define how the criteria should be applied. The following table provides some examples:

Example:

A) Required Service or Product - Importance for the business (in strategic terms and potential gains).

Is the project required to achieve business goals, meet legal compliance, or regulatory mandates?

1 = not required
9 = required or mandatory

B) Strategic Alignment

To what extent is the project aligned with our organization's overall strategies?
1 = does not align
5 = aligns with some strategies
9 = aligns with all strategies

C) Effort demanded

How much effort necessary to implement (Including men power, money and time)?

1 = Big effort

5 = some effort
9 = little effort

D) Impact in terms of telecom costs

How much direct impact in terms of telecom cost the project has (Savings or increase)?

1 = Increase the cost
5 = neutral
9 = Big savings

2. Establish criteria weight.

Place your criteria in descending order of importance and assign a weight. Note that when a project is scored, the numeric rating the project is given for a particular criteria is multiplied by the criteria's weight to create a priority score.

Weight examples:

Required Service or Product   :   Weight = 5
Strategic Alignment   :   Weight = 4
Effort demanded   :   Weight = 5
Impact in terms of telecom cost : Weight = 4

3. Create the matrix.

List your criteria down the left column and the weight and names of potential projects across the top

4. Work in teams to score projects.

Review each project and rate the project on each of the criteria. Next, multiply the rating for each criteria by its weight and record the weighted value. After evaluating the project against all of the criteria, add up the weighted values to determine the project's total score.

If participant numbers allow, it is helpful to work in teams. Working in teams can produce more objective results, since differing perspectives can be considered during the rating process, an external consultant maybe very helpful at this point.

It's always a good idea to go through the process with the whole group for a couple projects to help establish a common understanding of the process and to ensure a good comprehension of the criteria and their meaning.

5. Discuss results and prioritize your list.

After projects have been scored, it's time to have a general discussion to compare notes on results and develop a master list of prioritized projects that everyone agrees upon. Note that the rating scores are an excellent way to begin discussions, yet still allow room for adjustment as needed.

Remember that the prioritization matrix itself is just a tool, and the analysis of the ROIs of each initiative and even the identification of the possibility of gains through some initiative is a very hard task.

So we have three challenges:

1) The first one is to identify what could be done to improve the telecom infrastructure. That means the identification of all possible projects to be implemented. Here we have the demands from the business, the demands from regulatory or legal compliance and the initiatives linked with reducing the current expenditures and improving quality.
2) The second challenge is to associate an economic impact to each project, that means its implementation costs and its on-going costs (ROI).
3) The third is to classify properly each project. People scoring the projects should do so using their best judgment, criteria such as strategic alignment and effort demanded are very difficult do rate precisely. These evaluations are kind of subjective and demand a good understanding of the implications of what is being proposed and how it is going to be implemented.

Upon review, the evaluation group may decide that a project needs to move up or down in priority, despite the score it received. These types of adjustments are expected and help fine-tune the priority list. Be sure to vet the results with others in the organization, as well as with the business and others stakeholders.

It is advisable to review the projects with different hierarchical levels within the organization, for instance you may review each project with the technical teams after that with your senior management and only after all these discussions with the IT director and Business.

Example:

Here we are going to discuss a real telecom plan seen in practice how the priority matrix is used:

This priority matrix was developed when preparing the annual telecom plan for a large retail organization in Brazil whose expenditure with telecom spins around R$ 1,500,000.00 (USD 500,000.00) monthly. It was identified that five initiatives could be implemented and together they would be able to bring the current cost down 23%.

Note that in this particular example we are bypassing the whole process of identifying what could be done, which in itself demands a lot of knowledge and focusing exclusively in how the initiatives were ranked. In addition of that only one of the initiatives listed was related with the performance of the business the other four were almost entirely within the orbit of IT and telecom.

Project	Description	Required Service or Product	Strategic Alignment	Effort demanded	Impact in terms of telecom cost month
Backup links sites	Deployment of ADSL links as backups in the small size sites	No	No	Medium	R$ 100.000
Deployment of GSM gateways	Deployment of GSM gateways (Separation of On net calls and off net calls)	No	No	Medium	R$ 30.000
Call center operation	Adjustments in the mailings and diallers	Yes	Yes	High	R$ 100.000
Data links renegotiations	Providers rearrangements	No	No	Medium	R$ 20.000
Bill auditing reimbursements	Auditing and reimbursements from incoming calls	No	No	Low	R$ 81.500

The initial classification wasn't done (Numerical scale) for each criteria, just a more direct evaluation. Subsequently the classifications were replaced by numbers:

Project	Description	Required Service or Product	Strategic Alignment	Effort demanded	Impact in terms of telecom cost month
Backup links sites	Deployment of ADSL links as backups in the small size sites	1	1	5	9
Deployment of GSM gateways	Deployment of GSM gateways (Separation of On net calls and off net calls)	1	1	5	7
Call center operation	Adjustments in the maillings and diallers	1	5	1	9
Data links renegotiations	Providers rearrangements	1	1	5	6
Bill auditing reimbursements	Auditing and reimbusements from incoming calls	1	1	9	8

Next step we introduce the weights:

Required Service or Product   :   Weight = 5
Strategic Alignment   :   Weight = 4
Effort demanded   :   Weight = 5
Impact in terms of telecom cost   :   Weight = 4

Project	Description	Required Service or Product (weight 5)	Strategic Alignment (weight 4)	Effort demanded (weight 5)	Impact in terms of telecom cost month (weight 4)	Total	Ranking
Backup links sites	Deployment of ADSL links as backups in the small size sites	5	4	25	36	70	2
Deployment of GSM gateways	Deployment of GSM gateways (Separation of On net calls and off net calls)	5	4	25	28	62	4
Call center operation	Adjustments in the maillings and diallers	5	20	5	36	66	3
Data links renegotiations	Providers rearrangements	5	4	25	24	58	5
Bill auditing reimbursements	Auditing and reimbusements from incoming calls	5	4	45	32	86	1

That allowed us to rank the projects:

Ranking	Project	Description
1	Bill auditing reimbursements	Auditing and reimbusements from incoming calls
2	Backup links sites	Deployment of ADSL links as backups in the small size sites
3	Call center operation	Adjustments in the mailings and diallers
4	Deployment of GSM gateways	Deployment of GSM gateways (Separation of On net calls and off net calls)
5	Data links renegotiations	Providers rearrangements

This is the summary of the projects to be implemented during the next year and the priorities associated with each one.

6 Why Negotiating Telecom Contracts is Different?

Negotiating telecom prices well is difficult. It happens not only because we usually have monopolistic markets or telcos negotiate particularly hard. Negotiating prices of telecom services well is also difficult due the fact that when negotiating telecom services, we have not only to compare the prices per service, but also compare the prices of the different possible transport strategies.

Making an analogy with the physical world it would be like making quotations for cargo transportation where you have different modals (trains, Trunks, ships or airplanes) and you have to quote not only among the same modal. Following the analogy you will have to map the origins and destinations to verify which kind of company serves the location and you will have to check the volumes and requirements you have between each location in order to check which modal fits your needs. The same situation happens when negotiating telecom contracts.

Very often we see people executing this kind of work who doesn't understand this basic fact. Let's give some examples of what we mean by "comparing the prices of the different possible transport strategies" in the telecom world:

- Voice Services: The cost of the spoken minute has to be compared not only among the service providers' but also with the cost to transport them through the private voice network.
- Mobile services: The cost of the spoken minute fix-mobile cannot be compared only with the price offered by fixed trunks providers,

but also with the alternative of providing dedicated mobile trunks or gateways GSM installed in the PBXs.
- Data services: The cost of providing backup links to your sites through a MPLS network maybe a lot higher than providing it through a mobile G3 network.

To be able to speculate about transport strategies possibilities becomes necessary to understand very well the traffic. That applies to both voice and data. For instance it is important to map things such as if there is internal traffic (Ex: Among the organization own voice trunks or own mobiles), if there are particular point of traffic interest (Maybe you have some client or supplier with whom there is a large volume of traffic) or what is the volume between each area or country code.

Of course the underline assumption is the fact that you can map the different types of traffic and redirect it through different paths if you see advantage in doing so.

A subsidiary aspect of this approach (comparing the prices of the different possible transport strategies) is the fact that if you have the possibility of selecting the types of traffic and redirect it you may have the possibility of selecting the best alternatives of cost (Note that no necessarily tariffs) for each specific type of traffic. That frees you from negotiations where you have to select a sole provider for all types of services (Although all or nothing contracts may yield good discounts due volume aggregation).

Comparing and eventually adopting different transport strategies for different types of traffic is crucial to be able to negotiate well and archive good prices. If you adopt the traditional approach of simple comparison between prices your chances for driving down costs are limited.

To make things a bit more complicate providers usually associated a minimum value (or minimum usage) to a set of tariffs. Usually this correspondence is inversely proportional; the more volume you have the cheaper is the unitary price of the services. That brings you to a dilemma where if you redirect some types of traffic through other telco (or your private network) where your cost per service is lower the volume remaining maybe not enough to keep you within the best prices bracket.

In addition of all these factors there are also some risks which push to some strategic considerations (adding complexity to this process):

- The risk of concentrating too much business with few providers
- The risk of locking the organization into contracts without flexibility
- The risk of having contracts too long (three years is generally accepted as a good time span)

Of course we are talking about the ideal scenario, in most cases what we see are simple RFPs (Requests for proposals) where the resources/volumes today in use are listed together with a basic specification of the quality of services desired. Very rarely we see quotation processes preceded by detailed traffic analysis and good studies about the alternatives of transport available.

But the benefits are worth the effort. The fact that we are not just comparing the costs per service but also the costs per transport strategy is a fundamental point and does make a difference. Just to illustrate what we mean in practical terms our experience shows that a traditional negotial approach yields between 10 and 20% reductions against 30 and 60% when including in the analysis the possibility of using different transport strategies for different types of traffic. This can represent a huge amount of money in a large corporate network.

All these things are a way beyond of what can be done by typical purchasing departments and are the reasons why the telecom manager has to be part of any telecom negotiation.

Considering the complexities involved, it is advisable to start planning at least one year before the actual renegotiation. Of course, the time frame depends on the size and complexity of the network.

In the subsequent example we describe a case where the alternative of transporting the voice traffic through a private network was used as a price baseline to negotiate with the providers. This case took place in the Brazilian arm of a worldwide-diversified financial services company:

- The total of spoken minutes per month was 4,519,676
- The number of calls month was 1,396,327
- The average call duration was 3.23 minutes.
- All calls were handled by a call-center located in Rio de Janeiro city and originated from all parts of Brazil.
- Current contract with one of Brazil's main ILECs (OI) paying a flat rate.

The case depicts the renegotiation of the current contract whose expenditures was around US$ 828,000 per month.

In the end the negotial process generated savings of approximately US$ 270,000.00 per month bringing the cost down by 32%. The renegotiated structure would cost around US$ 558,000 per month.

The alternatives of providers available were EMBRATEL, OI, TELEFONICA and INTELIG/TIM (all of them providing dedicated and switched connections). In Brazil the calls within the same area code are charged and the 0800 calls can be charged differently depending on where they originated and the type of the service (mobile/fix).

The process encompassed two different steps: First involving direct comparison of proposals with the potential providers (**what we call traditional approach**) and the second involving re-design of the network structure including consideration of a private network and discussion of this project with the current provider. (**What we call "comparing the prices of adopting different transport strategies"**)

<u>First step</u>

This step consisted of comparing the company's current tariffs with the market alternatives available. We quoted the telephone bill (0800) verifying how much it would cost if it was charged using eight different proposals from four service providers. OI (basic plan and 31 empresarial plan), TELEFONICA DE SPANA (basic plan), INTELIG (two specific proposals) and EMBRATEL (basic plan and one specific proposal). These quotations gave us the view of what savings would be achievable through negotiation:

Charging plan	Value	Difference	%	Average price per minute
Telefonica de Spana Basic Plan	USD 2,234,604.00	USD 1,406,232.04	169.76%	USD 0.49
EMBRATEL basic plan	USD 2,073,483.77	USD 1,245,111.81	150.91%	USD 0.46
Teemar Basic plan	USD 2,064,738.00	USD 1,236,366.04	149.04%	USD 0.45
Teemar 31 Empresarial (RJ) plan	USD 1,064,075.22	USD 235,703.26	28.45%	USD 0.24
Telema r specific plan ourrently used	USD 828,371.96	USD 0.00	0.00%	USD 0.18
EMBRATEL specific proposal	USD 673,642.35	(USD 154,729.61)	-18.68%	USD 0.15
Intelig (1) specific proposal	USD 752,195.66	(USD 76,176.31)	-9.20%	USD 0.17
Intelig (2) specific proposal	USD 687,619.40	(USD 140,752.56)	-16.99%	USD 0.15

As can be seen in the spreadsheet above, through negotiating tariffs it was possible to achieve at maximum 17% discount over the current costs.

Second step (Re-designing the structure)

Although there were discounts achievable through traditional negotiations (Step 1), the current cost was also compared with the alternative of using a private voice network to transport the traffic. In order to make this analysis possible the traffic was analyzed identifying the origination and destination of the calls (Traffic matrix).

The optimized structure identified included 22 regional nodes (as shown below) connected to Rio de Janeiro through dedicated data circuits. It was considered adopting local numbers in each one of these 22 nodes. The 0800 number would work only outside these 22 areas. This structure would cost US$ 557,838 per month generating savings of 32% (US$ 270,162.00) over the current expenditure. The 22 nodes were as shown below:

#	Node Name	Area code	Total number of users associated to the node	Number of users local	Number of users between 50 and 100 Km	Number of users between 101 and 300 Km	Number of users between 301 and 700 Km	Number of users above 700 Km
1	BELO HORIZONTE	312	109,203.00	47,843.00			15,801.00	32,126.00	
2	JUIZ DE FORA	322	38,946.00	17,501.00			10,783.00	10,662.00	
3	UBERLANDIA	342	59,831.00	16,509.00			9,408.00	29,371.00	
4	MACEIO	822	52,363.00	20,072.00			2,699.00	29,266.00	
5	MANAUS	922	28,884.00	21,888.00			192.00	312.00	
6	FEIRA DE SANTANA	762	51,236.00	26,508.00	3,703.00		2,822.00	4,363.00	
7	ITABUNA	732	33,812.00	18,515.00			646.00	9,574.00	
8	SALVADOR	712	35,486.00	34,179.00			1,106.00		
9	FORTALEZA	862	86,730.00	66,795.00			1,174.00	7,682.00	
10	BRASILIA	612	108,283.00	69,901.00	18,785.00		2,849.00	2,686.00	
11	VITORIA	272	78,184.00	38,052.00			3,990.00	33,188.00	
12	GOIANA	622	82,807.00	52,717.00			5,888.00	12,435.00	
13	SAO LUIS	982	30,017.00	14,532.00			2,339.00	4,147.00	
14	CUIABA	653	46,452.00	23,131.00			92.00	3,032.00	
15	BELEM	912	22,597.00	18,196.00	288.00	334.00	146.00	0.00
16	JOAO PESSOA	822	36,580.00	24,787.00			4,658.00	7,063.00	0.00
17	RECIFE	812	102,397.00	90,704.00	3,236.00			8,457.00	
18	CURITIBA	412	96,423.00	38,524.00			7,380.00	23,711.00	2.00
19	RIO DE JANEIRO	21	113,873.00	86,782.00	7,465.00	1,172.00	19,484.00	138,636.00
20	NATAL	842	26,537.00	25,189.00				1,348.00	
21	PORTO ALEGRE	512	52,124.00	27,666.00	225.00	3,571.00	15,714.00	0.00
22	SAO PAULO	11	106,302.00	60,125.00			23,184.00	22,993.00	
	TOTAL		1,399,041.00	848,487.00	33,707.00	100,802.00	277,710.00	138,636.00
	Percentage		100.00%	60.65%	2.41%	7.18%	19.85%	9.91%

Others simulations were run identifying how much the structure would cost if deploying distributed IVRs and the local callers paid for the calls.

Backbone Cost	USD 116,904.7
Access Cost	USD 424,983.5
Hardware Cost	USD 15,950.0
Total:	USD 557,838.3

When confronted with the scenario where the organization would build its own structure, relaying on the telcos only for the local accesses, OI backed down and provided a better proposal where the cost per minute would bring the overall cost close to the US 550.000 monthly. Note that until this point the alternative would be to use INTELIG with a proposal which would bring the cost down by only 17%.

This is a classic example where the people negotiating thought not only in terms of how much discount could be achieved through comparing directly different providers but also in terms of the gains attainable through using a different transport strategy. We should notice that the best proposal achieved through the traditional process yielded savings of USD 170.000 monthly (step 1) compared with USD 270.000 when matching the transport costs to a private network (step 2). A difference between 17% and 32% (USD 100.000 month) no doubt is worth the effort.

7 Negotiating with the Service Providers

Negotiations with service providers are part of the functions of a telecommunications department in a large organization. In these day-to-day negotiations quotations, proposals and requests are made regularly, and technical and economical aspects are discussed. Negotiation skills are therefore assumed to be one of the basic skills of the typical telecommunications manager in a large organization. But even with experienced professionals, negotiating large corporate WANs is always challenging.

The negotiation of a large corporate WAN involves strategic aspects not present in the day-to-day negotiations. The financial numbers are high, the transition processes are lengthy, several aspects such as QoS are not easily mapped, and service providers are usually large corporations with experienced negotiators. All these reasons make proper planning essential.

In a typical organization, a complete renegotiation of the contracts supporting the WAN usually takes place at intervals not smaller than three years. Often, intervals even longer than three years elapse between contract reviews due to the rooted habit of just renewing existing contracts. This law of least effort of choosing to stay in the same contract even when it may not be the best alternative is one thing to overcome when negotiating. Regular negotiations are necessary to achieve the best results.

Considering the complexities involved, it is advisable that the IT or telecommunications manager faced with the need to renegotiate the organization's WAN must initiate the process at least one year before the actual renewal date. This is typically the time required to make a proper assessment of the current situation in terms of traffic per service, defining the strategy, identifying appropriate designs, and preparing a detailed RFP.

7.1 Considerations Prior to Negotiations

When negotiating a WAN structure, try to make all services contracts terminate at the same time, and, if possible, standardize the contract's lifespan. This will enable maximization of mix and size of services to be quoted/contracted. Frequently, it is not possible. However, this situation usually can be achieved within one or two cycles of the typical contract's lifespan. This standardization can also be achieved by extending existing contracts or making all new site services part of an existing bigger contract with a single co-terminus end date.

Frequently, competitive local exchange carriers (CLECs) and incumbent local exchange carriers (ILECs) in different countries own the entire range of services, from mobile telephony to providing leased line services utilizing their own infrastructure.

When the providers can offer an entire range of services, the telecommunications manager may manage to aggregate services increasing the overall size of the financial pot they are competing for. This strategy often makes possible the achievement of better prices.

7.2 Negotiation Strategy

Planning the strategy is crucial to successful negotiations. The process must consider several key factors that are typically determined by the makeup and culture of the organization. As an example, the telecommunications manager must evaluate the organization's disposition to manage and trust its providers. These are typically some of the factors to be evaluated when planning a WAN negotiation process:

- Agility required from the provider companies

- Organizational capabilities in different countries or regions of the globe
- Degree of centralization or decentralization of the IT infrastructure and decision-making
- Degree of telecommunications deregulation in countries of operation around the globe
- Language barriers
- Degree of local knowledge of telecommunications industry in other areas of the globe
- Ability or desire to manage multiple vendors and finance systems with different currencies, languages, cultures, and accounts payable environments
- Differences of control and capabilities in regions of the world over different technologies like voice, mobile voice, and data networking, which are often the result of an organization's historical growth and acquisition strategy
- Management's view of outsourcing

Note that the factors encompass aspects linked with the organization itself and its needs, the environments where the organization operates, and the providers available.

Control issues are another aspect that organizations usually ignore. Questioning and analyzing the existing control structures will either validate or clearly show that a strategy is not adequate to maximize use and efficiencies of technology tools. Sometimes, even the best technology is not suitable for a specific dysfunctional control structure. The people defining the negotiating strategy have to do such analysis.

As mentioned a strategy is clearly required when entering into service provider negotiations and it should address at least the following points:

- Clear statement of global approach to service providers, for example: One global provider (or one provider by country or region)

- Providers with local capabilities and infrastructure in areas of operation
- Best use of local providers in their areas of operation
- Local or a universal language capability in regions of the globe
- Finance capabilities, currency requirements, and payment terms
- Provisioning and sales cycle time frames
- Aggregation of all communication spend up for renewal by category with line items broken out for each of the following:
  - Outsource options
  - Voice mobile, private, fixed and public switched telephone network (PSTN)-based services
  - Conferencing services, including Web, voice, and video
  - Video streaming services
  - Remote access services
  - Data connectivity
  - High availability
- Clear internal understanding and communication regarding network capacity and QoS requirements
- Positioning the ideal topology based on flow analysis and geographical realities driven by political, organizational, and other regional influences
- Positioning at least two transport strategies
- Defining specific contractual elements up front, including:
  - Cancellation and penalty clauses
  - Accounts payable and dispute arrangements
  - SLA components
  - Penalties and contract cancellation conditions
  - Technology refresh options
  - Help desk performance metrics
- CPE ownership and management, which regional regulations might dictate
- Standard price lists

Considering the complexities involved with some of the topics listed, it is advisable to start planning at least one year before the actual renegotiation.

Of course, the time frame depends on the size and complexity of the network. It is our experience that the people executing this kind of work frequently display a lack of preparedness when negotiating a WAN. It is also very common to find people negotiating these structures whose view is purely economical. This view usually generates two very typical approaches:

- Kick-the-telco approach: - "There is no complication. We quote with all providers and just choose the cheaper one or force the one we want to get to the price we want".
- Minimalist approach: -"There is no need for detailed specification. Just send the providers the list of things we have today".

These views, although not entirely wrong, may work for a smaller infrastructure, but they do not consider a fundamental factor of this process. When negotiating a complete WAN, we are not only comparing the prices per service, but also comparing the prices of the different transport strategies. An example of comparison between different transport strategies is as follows, and it is the main problem with the kick-the-telco approach:

- Comparing the prices against the cost of a private voice network: The cost of the spoken minute has to be compared not only against the others service providers' prices but also with the cost to transport them through the private voice network.
- Mobile traffic: The cost of the spoken minute fix-mobile cannot be compared only with the fixed trunks providers, but also with the alternative of providing dedicated mobile trunks installed in the PBXs to transport fix-mobile traffic.

In addition to the above, the services contracted today may not reflect the actual organization's needs, which may have changed since the last negotiation assessment. This is the main problem with the minimalist approach. Therefore, when preparing to renegotiate the WAN, it is crucial is to accurately map the traffic, understanding perfectly from where to where it flows, the volume, quality requirements, and available transport strategies. In other words, before discussing technologies or contacting the potential providers, the telecommunications manager has to understand very well the organization's current and future needs in terms of traffic volume, interests, and profile.

The fact we are not just comparing the costs per service but also the costs per transport strategy is a fundamental point and does make a difference when negotiating with telcos. Besides the possibility of comparing different transport strategies, the only other alternative is the direct comparison between the same services. In this situation, each telco usually knows the limits of the others. The chances for driving down costs are limited. (This is usually associated with the volumes.) There is very little room for discounts. In this situation, there is a big chance that the current provider, who already has its investment paid for, offers the best price.

Large organizations may have several telecommunications providers. They usually concentrate their business with few of them. Typically, between one and three providers are responsible for 80 percent of all telecommunications expenditures. This strategy has the advantage of simplifying the operation and enhancing the relationship with the chosen providers.

However, we should be aware to avoid exaggerating this concentration of business. It may trigger overconfidence by the providers and loss of contact with other alternatives available in the market. I often hear comments like, "We don't want to have several providers." There is a fine line here, and no one should see having few providers as an end in itself. It is only worth as long it guarantees better prices, simplicity, and leverage.

In addition, when contracting services, it is always convenient to allow all providers to offer their services, even when they are not able to cover all network POPs. The basic logic must be, "Provider, quote your best price where you have coverage." Never put having few providers as a prerequisite when doing a quotation. This is an easy path for higher prices.

Adopt the logic of allowing all potential providers to present their best prices where they have coverage. There is no obligation to quote services for all POPs. This implies that you are not ruling out the possibility of having several providers. But it does not necessarily mean contracting in this way.

Eventually, a quotation can be done, allowing the providers to offer services only where they have coverage. Later, you may narrow down the options to the ones with wider coverage, using the cheapest prices of all proposals as negotiating references.

Although common, contracting only one service provider may not be advisable. Having at least two main providers may be better due to commercial and technical reasons. Commercial reasons are based on competition and the reduction in cost of moving services between existing suppliers, as opposed to bringing in a completely new service provider. From a relationship point of view, it is always good to create a situation where a service provider knows there is a concurrent competing contract to whom the organization can easily turn for its requirements. Technical reasons are based on security and risk mitigation. For example, it is advisable to use backup circuits contracted from a different provider than the one providing the primary circuits.

Of course, contracting a complete WAN, including the CPEs, with more than one provider may have its difficulties. A provider may have technical and/or commercial restrictions in allowing other providers to connect resources to its devices. This problem is rare though. In most cases, the quotations are done assuming that each provider will have its own CPEs, although this will vary between regions of the globe.

The management of multiple contracts may be seen as problematic, but, considering the fact that a typical telecommunications area already manages several contracts, although concentrating services into a few of them, this argument seems to lack substance. If the internal people are not capable to absorb this small amount of additional effort, something may be wrong with the way the organization manages its telecommunications contracts.

Segmentation may reduce volumes. Telecommunications managers may see this as tending to reduce the discounts offered. This argument, however, has several problems:

- This strategy does not eliminate the possibility of a provider offering all services in all areas (if it is capable to do so).
- This argument does not consider the already mentioned fact that we have to compare not only service providers and services, but also transport strategies. (One transport strategy may be applicable only in some areas.)
- Maybe more problematic, we may have potential providers whose prices are very good but limited to some specific geographical area. Depending from the percentage of our services within these areas, even if we pay a lot more for the services outside those areas, the savings could still be substantial.

With all these ideas in mind, we have to structure the quotation process. The quotation should be structured in such a way that the service provider must present a defined price per connection, and these prices must be associated to defined parameters. It enables visibility and transparency when contracting new services or changing/canceling existing ones during the contract life cycle, which will surely occur.

7.3 Pitfalls to Avoid

When preparing a RFP or a request for information (RFI), it is very important to avoid some basic pitfalls. I am going to list some of the more common ones:

- Price visibility: Very often, service providers are reluctant to present specific price lists for individual services and resources (insisting in providing a whole price). The difficulties associated with visibility of the telecommunications costs of new sites, including the difficulties linked with quickly verifying costs associated to a specific provider makes this strategy unacceptable.

- Gateway cost: A marketing practice prepares quotations of packet networks based only on the equipment and last miles costs. Such a quotation strategy generates a situation where the gateway cost has to be paid separately (just as another last mile). This charging strategy makes it difficult to clearly identify the connection cost per site and very often generates problems of over/under capacity for such gateways. The ideal is to avoid it when possible.

- Backup structure: When contracting backup circuits, the fact that it is not ideal to contract the main and backup circuits from the same provider should be given serious consideration. A single provider, even if the provider has two different physical infrastructures, will still expose the organization to administrative problems, which may affect the provider, such as strikes, bankruptcy, and so forth.

8 The Issue of Business Processes Outsourcing

The issue of BPO (Business Process Outsource) is indissociably from the fact that management capacity is a precious commodity in any organization. Most BPO functions are non IT but repetitive mundane tasks. That understood we can put in context the typical reasoning basing outsourcing processes. The typical arguments usually come like that: This is not our core business, the outsourcer share resources and therefore does it cheaper, outsourcing we will have specialized people looking after this subject. The underlying assumption of this reasoning is the fact that somehow, outsourcing, the organization will have the process better executed and/ or executed costing less.

In the context of the telecom area, outsourcing is always in the minds of the IT director, infrastructure manager and Telecom manager. The difference is just in the amplitude which this alternative is considered. The IT director dreams outsourcing the whole infrastructure, the infrastructure manager hopes to be able to outsource the whole telecom area and the telecom manager toys with the idea of outsourcing parts of the processes of treating bills and everything related with management of expenses and personnel.

Although outsourcing has been "in" and "out" of fashion and sometimes is sold/Bought as a panacea, the experience led us to have a more cautious approach to the pros and cons of this process. It is important to clarify that there is no point in being for or against outsourcing. As mentioned, practically all organizations, to some extent, adopt this approach for parts of its telecom processes.

The issue is to identify the reasons why it sometimes is not done in a way that works for both the BPO partner and the organization. The challenges usually come from five sources:

1) The processes have to be well structured in order to be properly outsourced. It can be done by the organization or by the outsourcer however, very often this preparation is not properly executed by neither.

2) Outsourcing a process sometimes leaves the impression that the company can "throw the process over the wall" in practice however there is still an important command and control function needed within the client organization, a lack of this oversight and partnership between BPO providers and client organizations very often leads to a failed outsource attempt.

3) Even outsourcing repetitive mundane tasks require a management oversight. In most organizations that often takes a secondary role, and is not always viewed as strategic and as a result do not always get the attention it deserves. However, failure to execute on these repetitive functions can have pretty serious consequences.

4) Sometimes there are conflicts of interests between the outsourcer and the organization, where if the outsourcer acts too efficiently it eliminates the own reason of his hiring (Ex: Billing auditors solving the root causes of billing errors).

5) The outsourcing process don't use shared resources and personnel (resources shared with other organizations) and therefore are exclusively dedicated to the client organization. Consequently the cost of these resources are equal or bigger than if they were company's own. That usually means that the professionals actually executing the processes earn less than if they were company's employees given the management cost layer added by the outsourcer. It usually means less qualified professionals.

That all said, it doesn't mean that outsourcing processes is too difficult to implement, it all depends of the process itself and how capable the organization is to deal with it. Here it is worth a parenthesis. Sometimes even if the BPO is not implemented in the ideal way it is already better than what the company was doing by itself and therefore worth the effort. Even when the outsourcing is more expensive, sometimes just the fact that the process now works justifies it. This point of view may sounds a bit defeatist, as if we assume that the company cannot change its ways and do better, but sometimes it is just how the reality is.

In general outsourcing tends to go better as the percentage of shared resources increase. For instance if you outsource your NOC to an organization which already has a big NOC serving several other clients there is a tendency that the service will be good (at least as good as you do internally) and the price will be compatible or lower than your current cost.

On the other hand the more dedicated the resources are the big the likeliness of having the same or worse standard when outsourcing. For instance if you have an internal bill processing team and decide to outsource it, the new outsourced professionals probably will work within your premises and although very likely not costing less will probably earn less than if they were your own employees. It usually means less qualified people doing the same job.

This problem may be minimized and somehow compensated by the fact that the outsourcer may assume the management effort of the process and may provide tools. If it does, you may have a better result even with less qualified professionals. Of course it may not happen and you can end up paying the same and having less.

Other important point that we have to keep in mind is the fact that when calculating the costs of keeping a process in house the costs are not only the ones associated directly with the process but all the cost associated with the areas giving support to it. Therefore, the costs associated recruiting and training people and the costs associated with the operation and maintenance of the facilities have to be included in the calculation.

Regards business processes outsourcing it is also useful to keep in mind an old said **"Processes should be designed by genius so well that even idiots can operate it not designed by idiots in such way that only genius can operate it"**. This said highlights an important issue, if you are thinking about outsourcing a process think first about how to design it in the best possible way, once you did that doing it in house or outsourced becomes easy.

In the end, the issue is to evaluate how each process outsourcing is to be implemented and to be able to see the advantages and disadvantages of each particular case as it applies to your organization's requirements and culture.

9 Inventory Control
Why it is Difficult?

In this chapter I am going to discuss why it is so difficult to create and maintain a good inventory control of the telecom resources.

Having a centralized and updated inventory of telecom resources is very important. It allows a consolidate view of the resources what in its turn is the basis for management initiatives. There is a general said which is very applicable to this specific context: **"you can only improve what you can measure"**. Having a complete view of what you have and how much it costs is the basis for everything. Nevertheless, it is very common to find large organizations where there is no centralized inventory of telecom resources (equipment and services),this happens due four main reasons:

The first reason why it happens has to do with the fact that these organizations frequently have several business units often operating in several niches and countries. That leds to a scenario where you often have several telecom teams operating with almost complete autonomy to cancel and contract services. Consequently, you have a situation where these decentralized telecom teams keep their own decentralized inventories of resources and control the insertion and remotion of resources to/from the inventory in a non-standardized way.

The second reason is linked with the fact that even when the telecom management is centralized it very often focuses exclusively in the technical aspects of the management of the telecom resources. That leds to a scenario where you do have a technical inventory of the resources but the linkage of this technical inventory with the administrative entities such as business units, contracts, resources costs, invoices and addresses is not done. That leds to a situation where the updates made by the technical teams have a loose linkage with what the telco actually do or with what actually is being charged for.

This issue of "exclusively technical inventory" creates all sorts of problems; the bigger one is the creation of two realities: You have the technical reality (what the organization actually uses) and financial reality (What the organization actually pays for). Over time the gap between both realities tends to widen, making the inventory unreliable.

Here we very often see a controversy. Some telecom managers argue that the fact that there is no linkage with the other administrative entities is not a big deal and the inventory managed in this way is not only useful but in fact quite correct because it represents what the organization is actually using. They often use the NOC as an example how things are well organized and every resource is well documented.

We don't believe this perspective is correct. It is our understanding that there is only one inventory and no gap should exist between the technical and the financial realities. However, if there is a gap, is the financial reality that counts. That means that doesn't matter if you are not using a resource, if you pay for it, it has to be in the inventory of resources, even only to allow you to cancel it and stop paying for it. That may sounds as a bit obvious statement but this perspective is very often lost by telecom managers with an exclusively technical mindset.

The third issue that makes keeping a good inventory control difficult is the problems associated with linking the control of the inventory with the processes of provisioning, changing and canceling resources. In other words how to control what enters, what changes and what leaves the inventory.

These processes have to be associated with the requests made to the service providers and should insure the control over some key milestones such as in the case of provisioning a new resource: 1) When the request was done, 2) When the resource was installed and 3) When the resource was accepted by the organization and therefore officially included in the inventory of active resources and the invoicing authorized.

In the case of canceling an existing resource: 1) When the request was done, 2) When the resource was canceled and 3) When the resource was checked as cancelled by the organization and therefore officially removed from the inventory of active resources and the invoicing blocked.

Or in the case of changing an existing resource: 1) When the request was done, 2) When the resource was changed and 3) When the resource was checked as changed by the organization and therefore officially changed in the inventory of active resources and the invoicing authorized.

The fourth and last issue is how to make a linkage between the inventory and the invoice processes. This is important not only to guarantee that you are not paying for resources which are not yours but to reflect the inclusions, cancels and changes made in the inventory in the bills, guaranteeing that the technical and financial realities are and remain the same.

In summary we have four types of problems: 1) make decentralized telecom teams to place their data in only one location in a standardized way. 2) Make the inventory that includes technical, financial and administrative information about the resources 3) Organize and standardize the processes of including, removing and changing the data of the inventory. 4) Guarantee linkage between the changes in the inventory and the process of paying the bills

Considering all these four issues we can speculate if it is possible to do a good inventory control in a large organization without a telecom resource management software. It is our understanding that although not impossible it is surely much more difficult. A good telecom resources management software can centralize the information (even in highly decentralized organizations), standardize data and procedures, unify the technical and financial realities, provide a platform where the interactions with the providers can happen and make a linkage between changes in the inventory and the process of paying the bills.

10 Asset and Service Inventory Management

As discussed in the previous chapter a crucial aspect when managing a wide area network is control the inventory of telecom resources. To have an effective control of the telecom costs it is basic to have precise information regards at least the following entities:

- The organizations and the business units served by the network;
- The physical locations (Addresses) touched by the network;
- The points of presence of the business units in the physical locations;
- The service providers supporting the organization;
- The contracts with each service provider including the price lists of each contract and the resources supported by each contract (Including the tariffs applied to each specific type of traffic);
- The equipment used by the organization including the ones whose ownership belongs to the organization and the ones rented or leased including their location and cost;
- The last miles used by the organization including information such as capacity, location and cost (The term last mile although more often used referring to data accesses is also applied to fix and mobile voice trunks);

- The dedicated connections used by the organization indicating the capacity and the location of each end (A and B);
- The control of all interactions with the service providers – requests of new resources, cancel, changes and maintenance of existing ones (Every interaction must be documented and traceable);
- The monthly cost of the infrastructure per organization, business unit, service, service provider, contract, address, point of presence and per individual resource (equipment, last mile and connection).

When stating that the managers must have this information it doesn't mean necessarily that they should have a system to control it, although more difficult such inventories and controls can be held in spreadsheets (telecom expense software are discussed further in this chapter). The key point is that if a manager of a telecommunications network cannot provide this information quickly and accurately that means he/she isn't managing the structure effectively.

Although it may sounds a bit strange, it is extremely common to find situations where the people who are supposed to manage the network don't have a complete view of the resources under their responsibility. It is very common scenario in large multinational companies where the structure is contracted by different groups within the organization (Ex; Part of the network is contracted by the headquarters and part by the local teams). To add to this problem it is very common to see different business units having autonomy to contract telecommunication resources, sometimes not even bothering to inform the people who are supposed to manage the whole structure. This often leads to situations when due to either management failures or cultural/systemic situations, manager's don't execute effectively their mandate.

It is important to notice that the existence of a NOC (Network operational center) and/or monitoring tools do not guarantee by itself the existence of an adequate inventory of resources and above all it doesn't guarantee an adequate link between resources contracted and values expended (expenses or depreciation). It is absolutely crucial to build and maintain a solid link between the physical and the financial aspects of the network.

10.1 The meaning of each controlled entity

Subsequently I am going to discuss the entities which should be controlled normalizing their meanings. The understanding of these concepts will allow a better understanding of the subsequent topics. Here it is interesting to notice that although most of these concepts may sound extremely simple and even obvious, when combined they may not be. In addition, we must understand them well enough to be able to recognize them even where the nomenclature is completely distinct. We divide the entities to be controlled into three types:

- Functional
- Technical
- Structural

Note that this classification is arbitrary and aims only to make the understanding easier. Our objective here is to give you a structured way to control your data, even if you use spreadsheets I strongly suggest you to organize your data as described here.

10.1.1 Functional

The entities classified, as functional are entities associated with the network but not components of the network itself. There are eight entities, which I classify as functional:

- **Organization:** Organization is the entity which encompasses the business units. A practical understanding of what that means can be visualized if we think about a state government as the organization and its several departments and state companies as its business units. The same concept can be applied to a large multinational corporation where its several country operations/business segments can be considered as independent business units;

- **Business unit:** Business unit is a subset of an organization whose criteria for segmentation can be multifold (geographical, type of activity etc.);
- **Addresses:** Addresses are all physical locations where the organization has telecom resources in other words all addresses touched by the organization's network. Note that the address may not belong to the organization, it may be a client's site for instance, still if there is a telecom resource in this location which belongs to the organization, we have a point of presence of the organization there, and consequently this address has to be identified.
- **Points of presence:** Points of presence are the physical locations of the business units in the addresses, note that the same address may have points of presence of more than one business unit;
- **Service providers:** These are the companies providing telecommunications services for the organization;
- **Contracts:** Contracts are the agreements between the organizations/business units and the service providers. Contracts have associated price lists;
- **Price list:** Each contract has its associated price list;
- **Requests:** Requests are interactions between the organizations/business units and the service providers. The organization can request the installation of new telecom resources, changes, cancelation or maintenance of the existing ones;

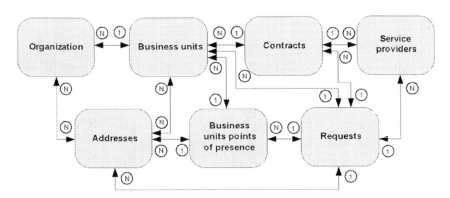

The relationships among the entities described are as follows:

- One organization may have several business units;
- One business unit belongs to only one organization but can be present in several addresses (points of presence);
- An address can host several business units (one point of presence of each business unit);
- A point of presence belongs to only one business unit and is located in only one address;
- A business unit can have several contracts with several service provides;
- A business unit can make several requests to several service providers;
- A service provider can have several contracts with several business units;
- A service provider can provide several last miles, equipment and connections for several organizations/business units;
- A contract can belong to only one service provider;
- A contract belong to only one organization;
- A contract can support several business units within the same organization;
- A request is linked to only one contract although we can have several requests associated to one contract;
- A request is associated to only one contract and consequently to only one service provider.

10.1.2 Technical

Technical entities are:

- Last miles (Data accesses, Internet accesses, mobile lines, voice trunks)
- Equipment - ports (CPE – Customer premises equipment, PBXs, mobile devices)
- Service provider network (represented by the cloud)
- Connections (PVC, CVP, Channels, traffic etc.)

The picture bellow represents the relationship among these four elements:

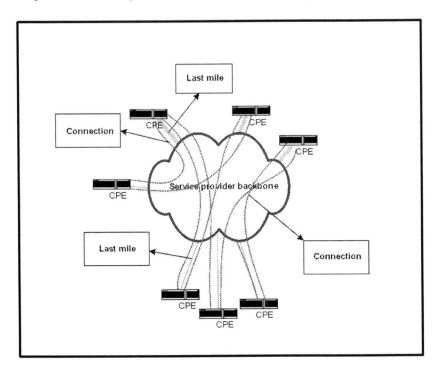

As we can see in the picture above, a connection between any given location A and any other location B is composed of six elements:

- End A equipment – with or without an associated port (CPE A);
- End A last mile, which is the local loop between the address A and the service provider backbone;
- Service provider backbone (represented as a cloud);
- End B last mile, which is the local loop between the address B and the service provider backbone;
- End B equipment - with or without an associated port (CPE B).
- The connection between A and B (usually called PVC or CVP);

Such elements belong to four types of technical entities: two last miles, two equipment, the cloud and the connection itself. Therefore we have here the four entities, which I call technical:

- **Last miles:** These are the local loops connecting the address to the service provider backbone (The provider backbone is usually represented as a cloud);
- **Equipment:** In data networks where the telco provides the equipment these equipment are also known as (CPE – Customer premisses equipment). These equipment are connected to the end of the last mile;
- **Cloud:** The entity "Cloud" do not constitute a real entity, in reality it is a simplification of the service provider infrastructure. The use of this simplification is necessary when analyzing conceptually a private network where the resources subcontracted to a service provider can be grouped and referred as "Cloud".
- **Connections:** Connections, as the name states, are connections between two physical locations (can be dedicated or switched) and are composed of five elements – Equip A, last mile A, backbone (cloud), last mile B, Equip B (The use of A and B terms doesn't mean any directional bias, the connections flow in both ways – usually);

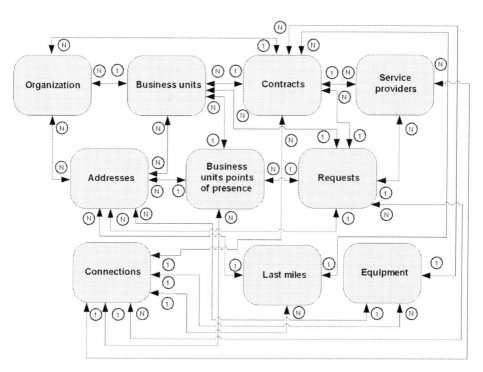

The relationships among the entities shown are as follows:

- A service provider can have several contracts each one with several last miles, equipment, ports and connections;
- A contract can include more than one last mile, equipment, port and connection
- A connection can be composed by last miles and equipment belonging to different contracts/service providers
- Each last mile, equipment, port and connection belong to only one contract
- A last mile and an equipment can be used by several connections
- A last mile connects to only one service provider backbone and to only one equipment located in only one address;
- A connection has only one equipment and only one last mile associated to each one of its ends;
- A connection is associated with only one point of presence (and address) in each one of its ends;

It is interesting to note that equipment and a last mile can be shared by different connections. It isn't uncommon to have equipment and last miles with one, two, three or more connections. Each connection though, is associated with only one equipment and only one last mile in each one of its ends.

Every connection encompasses eight cost factors (Equip A/Port A, Last mile A, backbone, last mile B, equip B/port B and the connection itself). Depending on the pricing strategy adopted by the service provider, some of these cost factors may seem to be zero. It is also possible that in some cases the equipment is owned by the organization, in these situations this equipment still has a "cost" which is represented by the TCO (total cost of ownership) which includes the WACC (Weighted Average Cost of Capital) of the organization plus maintenance and management costs.

The idea is to identify the cost of providing connectivity between two defined points A and B. Conceptually speaking every connection has these cost factors associated with the technical entities. The equipment and last miles costs are easily linked with the cost of providing connectivity between point A and B (although sometimes shared by more than one connection). The "Cloud" cost though is shared by all connections what makes it a bit more difficult to appropriate costs properly to a specific connection.

For clarification: The term "connections" refers to entities representing throughput between two points. Throughput has two basic attributes:

- Minimum guaranteed flow (Usually in packet networks referred as CIR);
- Maximum possible flow (Usually in packet networks referred as EIR).

This concept isn't linked to the technology used to provide the connection. In summary our understanding is that: The organization buys throughput among its points of presence from the service providers. In the picture we can see the eight entities which constitute a connection:

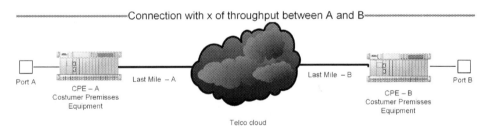

The eighth element not shown is the traffic. This is just the basic conceptual model and depending of the pricing model adopted by the service provider some items may not be present or may not have associated cost. We use this framework just to understand and normalize the several pricing strategies.

Example: A given telco may not charge for the ports and equipment, charging only for the Last mile and for the PVCs (the price varying with the bandwidth). Other telcos may charge for the last miles (the price varying with the bandwidth) and equipment (fix price) not charging anything for the PVCs. The point is that different services (and service providers) appropriate cost differently and this framework allows us to normalize this cost appropriation whatever it be.

10.1.3 Structural

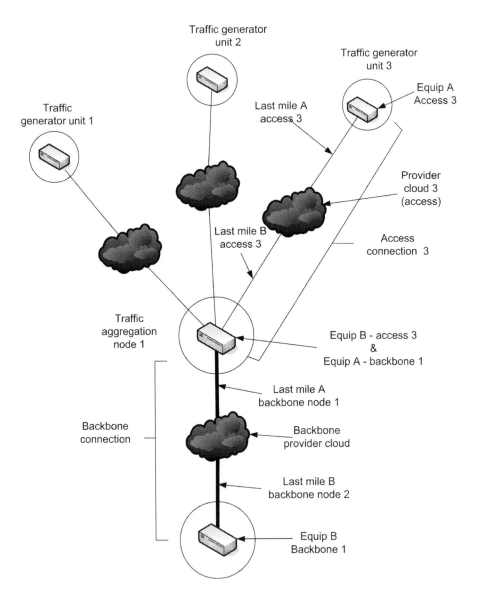

The picture 12 shows how the technical entities are articulated in a WAN where the topology isn't a star. This diagram is a bit more complicated than when we have a pure star topology. Here we have a situation where there are aggregation nodes and different service providers for the connections. In addition to that, we have connections linking the sites to the nodes (Access) and connections linking the nodes between themselves (backbone). This picture helps us to become familiar with the entities which I call structural:

- Traffic generator unit: In this context (corporate network), traffic generator units usually are workstations and people (Although some variation of this basic concept may exist)
- Traffic unit: The traffic units are the groups of traffic generator units, in a corporate network these units are usually the organization's points of presence, in a context of a call-center these units are the call-center users grouped by area code.
- Traffic aggregation node: Traffic aggregation node is where (for transportation reasons or for final destination) we aggregate the traffic before distributing it. Following an analogy with a logistic network the nodes would be our warehouses where the goods are stored and from where they are distributed. Here is interesting to understand that a given location can be a traffic unit and a traffic aggregation node at the same time.
- Backbone: The term backbone in this book refers to the connections between the traffic aggregation nodes.
- Access: The term access refers to the connections between the traffic units (sites) and the traffic aggregation nodes (nodes).

It is important to mention that the term backbone isn't associated to the bandwidth (Although there is a tendency for the backbone connections to be bigger).

Another important aspect, not always obvious, is the fact that equipment and last miles can be part of access and backbone connections. (Ex. Equip node 1 picture 12 supports connections to the node and to the sites).

It is important to understand that such concepts are applicable both to private and public networks (dedicated or switched). In this book our focus

Luiz Augusto de Carvalho

is in private ones and therefore some concepts expressed may have their meaning slightly changed when applied to public networks.

The subsequent pictures illustrate how the traffic generator units are associated to the traffic units. As we can see, in a pure data network the traffic generators units are devices like workstations, PDVs and servers, in a pure voice network the traffic generators units are people (or extensions) and in an integrated one, both.

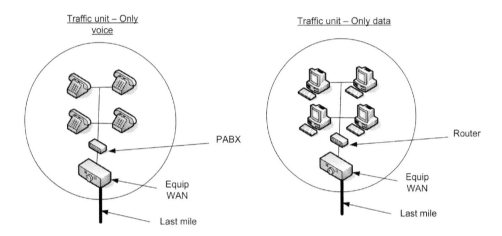

10.2 How to map the information

When building an inventory of telecom resources, besides understanding the meaning of the controlled entities, it is very important to map the information following a rational sequence. The following of this sequence is important because each layer of information supports the subsequent one. Therefore it is very important to organize de data to be gathered in a predefined sequence as follows:

1) Organization
2) Business units
3) Addresses
4) Points of presence
5) Service providers
6) Contracts and price lists
7) Last miles and equipment
8) Connections

In the previous item of this document I defined the meaning of each entity now I am going to show that why when executing the data gathering to populate the databases it makes sense to proceed in sequence building layer over layer of information. Now we are going to re-discuss the meanings with emphasis in how to build a proper inventory of resources:

Organization: Organization is the entity which encompasses the business units this a quite straight forward concept and it isn't worth to re-discuss.

Business unit: As explained in the previous item a Business unit is a subset of an organization whose criteria for segmentation can be multifold (geographical, type of activity conducted etc.);

Note that once you have identified clearly the organization and de business units it is time to identify the addresses and points of presence. These two concepts may sound equal but in fact they are not. An address can have points of presence of several business units. For instance if we are talking about a large organization with two business units, you may have an address where both business units are present and therefore we have two points of presence within a single address.

Other important aspect to be considered is the fact that the objective is to map the location of the telecom resources, therefore we will have situations where the resource is installed within a client or a provider address, in these situations the client's address has to be identified and it will be considered as a point of presence of the organization within the client/or provider. In other words addresses are all locations where the organization has telecom resources. Having telecom resources meaning: Telecom resources whose financial burden is supported by the organization (even when installed outside its premises).

As already mentioned, it is highly advisable to follow a rational sequence when mapping the infrastructure, it happens because each layer of information gathered helps to build the subsequent one. For instance if we map the addresses properly we cannot fail do ask which telecom resource is installed in each address (identifying things such as "Why this site doesn't have a PBX?). Proceeding in layers like that helps to ask the right questions and avoid mistakes.

In terms of organizing the telecom contracts the sequence recommended is as follows:

- Service providers
- Contracts -> Price Lists

Service providers: As explained previously these are the companies providing telecom services for the organization;

Contracts: Contracts are the agreements between the organizations/ business units and the service providers. In general, we usually consider each bill as a consequence of each contract. That means if you receive ten bills per month it is as if you had ten contracts. Of course you may have only one real contract to which all ten bills are associated, however in terms of building a database of your telecom resources would be advisable to match the number of bills with the number of contracts (Even if they are not exactly correlated). The reason for that is because somehow you must be able to identify the number of bills you have to pay each month and if any bill doesn't appear you must be able to take notice.

Each contract must have an associated price list and each price list may have several items which must be classified as one of the following types:

- Last miles
- Equipment
- Ports
- Traffic
- Cloud
- Connection

To be able to properly document the price list it is very important to understand the rules applied when defining the transport prices. This understanding implies identifying the parameters and factors that the service providers use to determine their prices and correlate them to the six predefined possibilities described above.

Be able to put all contracts in the same framework is as important as it is trick. Important because it is what makes you able to identify your price structure throughout a multitude of services, service providers and countries and trick because the charging strategies sometimes are very opaque. In general telecom charging strategies oscillated between the following two extremes:

- 100% predefined values per services,
- 100 % variable values (depending from actual usage-for example: spoken minutes)

Although most services are charged somewhere in between these two extremes, data access usually has its price 100% predefined and usually have five cost components plus taxes and discounts:

- Last miles usually have their prices defined based on the nominal bandwidth. – Associated to "Last mile" in our framework;

- The maximum amount of ports and type of services supported usually define the price of the equipment. – Associated to "Equipment" in our framework.
- Ports usually have their prices defined based on the speed (bandwidth) – We may bundle the port costs into the equipment cost or associate it separately as "Port";
- Management is usually a defined value for the network as a whole. This value is sometimes divided by the number of circuits and charged on a circuit basis. – Usually we bundle this cost into de equipment or into the last mile cost;
- The connections prices (regardless PVCs, CVPs or point to point) usually are defined in terms of CIR, EIR, or nominal bandwidth. – Associated as "connection" in our framework;
- Taxes vary based on the country, state, and even city where the connection is installed;
- Special discounts based on the specific negotiations.

These cost components sum to the total cost for data access service: usually providing data connectivity between two points implies in the following cost structure:

Port A + Equipment A + Last mile A +(PVC/connection/cloud) + last mile B + equipment B + ports B) = Total cost plus taxes and discounts. Considering management bundled into the last mile or equipment cost.

Voice services usually are priced in a mix between 100% predefined and 100% variable and usually has only three cost components defining the cost:

- Last mile is usually a monthly subscription fee – Associated to "Last mile" in our framework – 100% predefined;
- Usage is usually the value charged that is associated to a volume of spoken minutes. The cost of the spoken minutes may vary depending on the distance, hour of the day, area code, national/ state borders, or day of the week. It is also common to find situations where a minimum amount of traffic is charged, regardless if it is

actually used or not (sometimes we have charging strategies in a per call basis- regardless the duration). This cost item is associated as "traffic" in our framework and we may identify the several types of traffic for example: Long distance, local, mobile etc. In our price list we identify the unit and the cost per unit. Ex: minute of long distance call – USD 0.01.

- Taxes normally vary depending from the country, state, or city where the connections are installed.
- Special discounts based on the specific negotiations

The formula for the total cost of voice services is therefore the sum of only three components:

$$\text{Last mile A} + \text{volume used} \times \text{cost per unit} =$$
$$\text{Total cost plus taxes and discounts.}$$

Although we may find variations of these basic components, in a vast majority of the cases, the prices will be defined based on this cost structure. Knowing how the tariff is structured is important for the understanding how the total cost is built.

Sometimes, the cost appropriation does not use all cost components. Often, this cost appropriation is divided between the two ends of the connection. For example, in a given phone call, the call originator usually pays for the subscription of his own trunk (last mile) and for the traffic. The user receiving the call also pays for the subscription of his own trunk. In a situation like that, the match with the cost components described previously is as follows:

- Equipment and port A: Cost appropriation is zero, but this may not be true in some cases, such as when the telco provides the PBX (called CENTREX services).
- Last mile A: This is the trunk subscription, which the call originator (caller) pays.
- Cloud – the telephone network: The cost appropriation is zero.
- Last mile B: This is the trunk subscription, the receiver pay for it.

- Equipment and port B: The cost appropriation is usually zero, but this may not be true in some cases, such as when the telco provides the PBX.
- Connection and traffic: The caller usually pays for the call based on the duration.

11 The Interaction with the Providers a Big Source of Inefficiency

The issue here is to discuss an important aspect of how people actually work in a typical telecom area, the issue of the interaction with the providers.

The typical telecom manager talks and exchanges e-mails with the service providers all the time. The same happens with all people within the telecom area. We evaluate that something around 40% of the time of a typical telecom team is spent interacting with service providers and these interactions usually lack efficiency. When we say "lack of efficiency" we need to specify better what we mean:

1) Things are said but not written, that leds to situations where sometimes you have difficulties to identify what was in fact requested and agreed, when and by whom.

2) Things stay with only one person, this person is the only one who knows what is going on regards specific issues. A big problem when the person get sick or goes away on vacation or leave the organization

3) Many people ask things to the providers through several means (phone, e-mail, telco provisioning system, etc.), sometimes duplicating requests or worse duplicating with smalls differences, that generates lots of calls and e-mails to patch things up.

4) Many interactions are demanded regards the same issue due the fact that most times the initial request (Whatever the means in which it was done) doesn't have all necessary information to

execute what was asked. That demands a lot of duplicate work (Lots of calls or e-mails to complement details that supposed to be informed in the first request)

5) People who doesn't supposed to be able to ask things to the telco directly end up doing so what adds confusion to an often already tense relationship

6) The interactions don't generate follow troughs, that means for example even when a team member manages to navigate through the whole process of installing a new resource there is no automatic linkage with the inventory of resources or with the billing process what leaves the billing team on the dark.

The result of these problems is a typical scenario where everybody seems very busy but when observing closely we perceive that much of what is being done are activities derived from not doing right from start. Very often we see interactions suffering from one or more of the six problems mentioned previously.

To help us think about the issue lets structure the interactions by type and by phase within the type. In general the interactions can be divided basically into five types:

1. Requests for New resources (and the whole process associated with quotation, evaluation, installation, test and acceptance).

2. Requests for changes in existing resources (Including the whole process associated with quotation, evaluation, installation, test and acceptance).

3. Requests to cancel existing resources (Including the whole process of deactivating the resources)

4. Requests for maintenance in the existing resources, including the whole process associated with test and acceptance

5. Requests for verification regards billing

If we think about these five types we can see that the first three of these types have five possible phases:

1. Organization is requesting quotation, feasibility study and deliver time to the service provider
2. Service provider answering what was requested by the organization (or informing that cannot attend)
3. Organization approving, negotiating or declining what was proposed: price and deliver time.
4. Service provider informing that what was requested was executed
5. The organization informing that the resource was checked and accepted or there is some problem.

The type four has three phases:

1. Organization is requesting maintenance in some resource to the service provider
2. Service provider informing that what was requested was executed
3. The organization informing that the resource was tested but there is still a problem (or everything is ok).

The type five is quite similar to type four

1. Organization is complaining about some billing problem spotted in some bill to the service provider
2. Service provider informing that what was complained was checked (accepted or not)
3. The organization informing that they agree or disagree with what the provider informs and arguing what it believes is correct.

At any given time, take a snapshot of what people is talking through the phone or writing through e-mails and 99% will fit into some of these categories and phases.

Once we understood the types of interactions and the phases, becomes easier to identify where the typical problems occur.

It is our view that the main and first problem is linked with not sending all the relevant information regards the request from the start. It usually generates several e-mails or calls from the service provider asking for details such as where precisely the resource has to be installed, who suppose to be the local contact, what exactly is going to be installed and so forward.

The second big source of inefficiency is linked with the difficulties to share information. The fact that frequently only one person knows what is going on regards specific process. That generates situations where if the person who did the request is away (on vacation, sick or whatever the reason) and the provider contacts the organization inquiring for details (which supposed to be forwarded from the start) no one knows de necessary details (or at least take some time to unearth the process – if there are shared written records).

If the person who had the details did not copied anybody or if there is no centralized e-mail group, nobody will know the details and the alternatives usually are call the person, do nothing, or worse start anew (giving complete new instructions to the provider)-what will generate more confusion when the person comes back and the additional instructions are not shared with him. None of these scenarios are good.

The third big source of problems is the fact that the complementary information is not added to the request as it flows from phase to phase. It is a difficult thing to solve because very often the teams involved are not the same. For instance, even if the telco informs that a resource was installed, the people within the telecom area responsible for testing it are not the same people who generated the request, and they may fail to communicate. Another example of this kind of problem occurs, when a request is concluded but nobody includes the data regards the equipment and last miles into the inventory – or share it with the billing team. Things like that.

To minimize these problems the key is to understand that there are five milestones which must be formal. These formal milestones must include a set of basic information which has to be informed to the service providers (3 phases – 1,3 and 5) by the organization and a set of basic information which has to be informed to the organization by the service provider (2 phases – 2 and 4):

1) The organization has to send a formal (written) request to the provider, specifying precisely what it wants, the location of the resource (including details such the person to be contacted in the site) and the contract to which the resource is associated.

2) The service provider has to answer formally if it can execute what was requested and if it can the timespan necessary to deliver (This answer may or may not include price given the fact that most requests are made within existing contracts) – This answer is often referred as **feasibility study**.

3) The organization has to formally accept or decline the timespan and the price proposed. If it accepts the clock starts ticking for the execution of what was requested (Installation of new resource, cancel or change of an existing one).

4) When the service provider concludes the execution it has to formally inform the organization about the conclusion, indicating all information (technical and administrative) regards the resource (Including things such telco ID, IP, Serial number and so on).

5) When the organization is informed about the fact that the provider executed what was requested (Or at least informed the organization it did), the organization has to test and formally accept (or not) the conclusion of the request.

Regardless all the talk or exchange of e-mails which may happen during the execution of a request, these five milestones should exist and be formal (written). This is what we could call "the basics". If you adopt a good form and a good process half of all interactions become unnecessary and productivity increases enormously.

Keep in mind that Telcos tend to be very bad at managing the whole request to delivery exercise, so you need to have them carefully tracked, particularly if there are dependencies within requests. The Telco side phases often slip, particularly when they are dealing with 3rd party providers (Most often when dealing with global requests in some less developed countries). Therefore, it is imperative to have everything documented and get status reviews on a regular basis.

12 Service Ordering & Changing control

In this chapter we are going to discuss the processes associated to request new resources and cancel or change existing ones.

Here we have to understand that the process of requesting a new resource usually doesn't occur through a retail negotiation, but is done within an existing contract where the price is already predefined. That means that what we are going to discuss in this chapter is more linked with the mechanics of requesting new resources and canceling or changing the existing ones not with quotations.

The definition of these mechanics is very important because it has a direct impact in the processes of controlling the resources inventory, checking the bills and controlling the contracts. To be able to affirm that a resource should be paid, you have to know who asked for it, when it was installed and how much it supposed to cost (which contract and which item in the price list it is connected to).

Besides, considering the fact that a new resource implies in additional costs (and a canceled one in a reduction of charging values) it is absolutely critical that each request made to the service provider must be formal and controlled.

The requests of new resources, cancel and changes in existent ones need to follow a predefined process. This process must include the following phases:

Formal initial request: This phase implies that the organization asked one of the three options (new, cancel or change) to a service provider and the service provider must answer this request within a defined timespan. Note that answer here doesn't mean actually do what was asked but inform if what was asked is feasible, within which time and how much it would cost. We generally call this phase – "Waiting service provider answer". Note that the quotation of how much the resource would cost constitutes what we call "retail negotiation" and unless you already have the item in your price list of an existing contract you have to do that for each resource requested.

Here it is worth mentioning that in some organizations the quotation process is separated from the actual ordering. In this scenario the order of a new service only happens after a quotation with many providers already happened. In this context the formal initial request has only the objective to define the deliver time of the resource. This strategy implies that nothing is asked to the provider until the price is already known and a contract already exists.

Formal answer from the service provider: This phase encompasses the answer of the service provider informing if it can execute what was asked, how long it would take and how much it would cost. Then the process turns to "Waiting organization approval".

Approval or not by the organization: Once the service provider informs if it can execute the request, within which timespan and how much it would cost, the organization evaluates if it still wants what was requested, maybe the cost is too high or the timespan too long. If the organization doesn't approve the request it is finished. If the organization approves it the request changes phase becoming – "Waiting service provider implementation".

Note that you have to have some mechanism to control the time between each phase, this time is usually set in the contracts as part of the Service level agreement – SLA. Therefore, if for example you have five days to the service provider do answer and five days to the organization approve or not, you have to have some control about the request status. If they were answered within the defined time or not, the same control is necessary to the time defined to the organization to approve or not the request.

Of course you must be able to control the schedule given by the service provider to execute what was asked. Therefore if the service provider for example answered that it can install a new last mile within twenty days, you have to have a mechanism to follow the request indicating if it is delayed or not.

Service providers informs that the request was concluded: When the service provider finishes what was requested it has to inform it to the organization. Once it was informed the request changes phase again turning to = "Waiting organization to test".

Organization tests if the request was fulfilled: The organization has to have a deadline to test and accept or not that the request was executed. If the request is accepted the request finishes and the financial implications become active. If the organization doesn't accept the request it returns to the previous phase –"Waiting service provider implementation". Here it is worth including two cavets: 1) usually there is a mechanism where if the organization defaults the time to test the request is considered accepted. It works as if the organization had waived the right to test 2) The request can return to the previous phase as many times as necessary, until the organization considers that what was requested was correctly delivered.

When we mention that the financial implications of the request were activated that means the cost of the new or changed resource should be automatically included in the contract or the cost of the canceled resource should be removed from the contract (and consequently from the bill).

The four phases described apply for requests of new resources and for changes and canceling of the existing ones.

Requesting maintenance for a telecom resource to a service provider requires a specific workflow. The main difference between the maintenance workflow and the general requests (new, cancel or change), described in the previous paragraphs, is the fact that in a request for maintenance there is no need for the service provider to inform if it can execute the request, the timespan and the price. All three information are predefined. That means in a request for maintenance everything works as if the first status was "Waiting service provider implementation".

Summarizing requesting services from the providers demand a set of rules and procedures. Basically: following the sequence:

1. The organization makes a request;
2. The service provider answer how much is going to cost and how long it is going to take to execute (can also say it can't execute the request);
3. The organization approves or not the price and the time;
4. If the organization approves, the service provider starts executing what was requested and the time countdown starts;
5. When the service providers finishes it informs the organization that the request was concluded and is ready to test;
6. Then, the organization checks if what was requested was properly executed and approves or not it.

If the request is for maintenance, the sequence is slightly different, given the fact that some phases are not necessary. In a maintenance request the organization doesn't have to approve the execution of the request and the service provider doesn't have to inform the time to execute it (it is defined beforehand in the contract as part of the SLA).

The exception for the rule described before will be the maintenance request for equipment without maintenance contract, situation where all steps have to be followed.

A request is the stage where the several actors of the process of requesting and executing the four types of tasks interact. Therefore, becomes necessary that each one of these actors edit the request as it goes through the process. Therefore, the service provider has to be able to enter the date when he/she believes the request will be concluded, then, the organization approves it or not, then, if it was approved, the service provider informs that the service is ready and so the process goes. As can be seen, when we talk here about

"Editing" a request the meaning is more "filling" it as the process goes. Each actor has to fill its respective fields as the process goes.

Note that to be able to link these actors phase by phase the deployment of tools are very helpful. Of course you may be able to manage this process without a telecom manage tool, but as we can see is very hard to follow all these steps and phases using for example Excel spreadsheets and e-mails.

The request process (Not including here the maintenance) has four phases and two possible statuses in each phase. (on-time and delayed). The maintenance request process has only two phases and two status (on-time and delayed).

A request for new resource or a request for change or cancel of an existing one can have the following phases/status:

• Waiting time definition by the service provider	On-time	Phase 1
• Waiting time definition by the service provider	Delayed	Phase 1
• Waiting confirmation by the organization	On-time	Phase 2
• Waiting confirmation by the organization	Delayed	Phase 2
• Waiting - execution by the service provider	On-time	Phase 3
• Waiting - execution by the service provider	Delayed	Phase 3
• Waiting test/approval by the organization	On-time	Phase 4
• Waiting test/approval by the organization	Delayed	Phase 4

This sequence presupposes that there is a service level agreement (SLA) in place defining the four time limits, which separate on-time from delayed:

- Time limit – 01 – Maximum time that the service provider has to inform the organization how long it is going to take to execute the request "Waiting time definition by the service provider"
- Time limit – 02 – Maximum time that the organization has to confirm or not the request based on the "time to execute" informed by the service provider "Waiting confirmation by the organization"

- Time limit – 03 – Maximum time to execute the request, informed by the service provider when it answered the request, and approved by the organization when it confirmed the request "Waiting - execution by the service provider"
- Time limit – 04 – Maximum time that the organization has to test and approve or not the service executed once the service provide inform that the service is ready. "Waiting - Test/ approval by the organization"

Once the request is generated by the organization, the service provider has a given number of days to answer it informing how many days will be necessary to execute the request (or to inform that it is not possible to execute what was requested). The number of days to answer is defined in the contract.

Once the "time to execute" was informed by the service provider, the organization has a given number of days to confirm or not the request. If you don't accept the time proposed by the service provider to execute the service the request dies, but you still have to inform it that you did not accept.

Approving a request has economical and operational impacts and therefore only authorized people should be able to do that.

If the organization accepted the "time to execute" proposed and approved the request, the time for installation starts counting. The service provider will be evaluated if it succeed to deliver the service within the time of its own defining. Therefore, in this context "Delayed" means that the service provider failed to deliver the request within the time, which was informed to and agreed by the organization.

Once the execution of the request was concluded, the service provider must inform the organization (including all details about the resource Ex: in case of installing a new link: Circuit ID, IP, DLCI etc). Once informed, the organization has a given number of days to test and accept or not the service. The number of days to test and accept or not, should be defined in the contract.

If the organization tests the service executed and do not approve it, the request returns to the status "Waiting - execution by the service provider".

If the organization do not test the service executed and don't approve it within the time defined, the request stays as "Waiting test/approval by the organization" but "delayed". Sometimes it is agreed with the service provider that resources not tested within a given number of days are considered accepted by default.

In its turn maintenance requests don't have to be answered or approved they go directly to the "Waiting - execution by the service provider":

1) Waiting - execution by the service provider   On-time   Phase 3
2) Waiting - execution by the service provider   Delayed   Phase 3
3) Waiting test/approval by the organization   On-time   Phase 4
4) Waiting test/approval by the organization   Delayed   Phase 4

As already mentioned the maintenance request doesn't have the phase 1 and 2 given the fact that it isn't necessary to a maintenance request to have the time to execute informed by the service provider and subsequently having this time approved by the organization. A maintenance request has this "time to execution" previously defined in contract. The number of hours to solve the problem is defined in the contract.

Basically a maintenance request, once made, demands immediate action from the service provider. The service provider has only to inform when it solved the problem, in order to allow the organization to test it. Then, accept that the problem was solved or not.

12.1 Unifying the communication channels with the providers

Usually, in large organizations, several people interact with their telecom providers, asking for new resources, canceling services, changing existing resources and asking for maintenance. This way of working although usual tends to generate several problems:

1) It is difficult to know what was requested to the providers, when and for whom. This happens because the requests and its answers from the providers stay restrict to the individuals who executed the request. Usually there isn't a common depository where everything asked to the providers and everything answered by them is stored and become accessible to all telecom team.

2) It is difficult to coordinate parts of the processes which very often are executed by different groups within the telecom area. Ex: A group of people is responsible for making the requests of new resources, a second group is responsible for controlling the provisioning, testing and acceptance of resources and a third is responsible for checking the invoices.

3) It is difficult to guarantee the quality of the requests given the fact that there is no standardized request form which makes sure that all necessary information are in fact properly and timely forwarded to the providers. Ex Complete address, site contact, contract and details of the resource.

4) It is difficult to guarantee that the communication channel is the right one what often makes a request to be forwarded to the wrong people within the providers. That may generate situations where the responsibility for problems become diffuse.

5) The registering of the requests (mostly maintenance requests) is made solely by the provider, in addition each provider has its own strategy to be accessed (portal, e-mail, call-center) in case of disputes the discussion will be entirely based on the provider records, what isn't ideal from the organization point of view.

6) It is difficult link parts of the processes. Ex: How to guarantee that a resource be paid only after tested and accepted?

A good telecom resources management tool can help with all these points. A system like that can store the requests from the organization and the providers answers in an organized way and even can be the platform over which the requests themselves are made and answered. Usually we should define three institutional channels and define that all communications between the organization and its providers should flow through them. Only one for the organization and two for the service providers (commercial and technical).

The process of interacting with the providers encompasses two phases

- Creation of requests;
- Following the requests.

The interaction with the service providers demand a flow of information which can be basically divided into the following sequence:

1. The organization makes a request;
2. The provider answers if it can or can't execute what was requested and if yes within which timeframe (feasibility study);
3. The organization approves or not the timeframe (if it doesn't the requests dies);
4. If the organization approves the request, the provider starts the execution and the delivery time starts counting;
5. The provider finishes executing what was requested and informs the organization.
6. The organization verifies if what was request was in fact executed and acknowledge (or not) the request as concluded.

Maintenance requests go from phase 1 to 5 directly given the fact the time to execute is predefined in contract and there is no need to re-approve the answer from the provider.

Summarizing: It is imperative to adjust the internal procedures in order to make feasible the implementation of a unified and standardized process of interacting with the service providers. The implementation of this process not only improves the quality of the interactions but also reduces the time demanded to execute and manage them.

In addition, a unified process of requesting services tends to reduce the dependence from individuals putting the process in more institutionalized level where all information is shared with the whole team and where the communication between the organization and the providers is not personal.

Therefore, the crucial point is to establish what we call the "institutional channels". A good telecom resources management system can be very helpful when comes to implementing this process.

At this point it is worth emphasizing that the idea of "institutional channels" is applicable in both ways, the organization sends requests through only one channel to only one recipient(or two if we adopt the separation of commercial and technical channels) and the provider also answers back through this unique way.

If there is the possibility, the provider may answer the request directly using the organization own platform. It isn't common but where possible should be tried.

Although it may sounds a bit obvious it is important to specify a concept: When talking about "institutional channels" we mean e-mails or phone numbers which are not associated to a specific person like an account manager.

This way of working "Through institutional channels" although sometimes a bit difficult to implement, once established tends to increase productivity, simplifies the process and makes easy to document the interactions between the organizations. In addition of that this strategy guarantees continuity, regardless of changes in personnel in the organizations.

To implement institutional channels we should be able to negotiate it with the providers and implement internal mechanisms to forward the requests though a defined channel. Here it is important to mention the fact that the establishment of defined institutional channels also helps enhance the security of the process guaranteeing that only what was asked through the institutional channels is valid. A common problem is having multiple people asking for resources on behalf of the organization.

13 How to make a Quick Assessment of the Current Scenario

Assess the current situation of the management of the telecom infrastructure in a large organization is no easy task (level of maturity included but not limited to it). This problem is made worse when we think that most people doing these assessments do so with time constrains and sometimes not counting with the collaboration of the people directly involved (Who usually doesn't like to see their performance evaluated).

The ability to do these quick assessments is crucial in several circumstances such as a new IT manager taking over, a telecom manager deciding if he/she changes his job or a telecom consultant preparing a proposal of services. In all these scenarios (And in many others) you have to be able to quickly evaluate the current scenario, identify the main challenges and of course measure the effort demanded to face those scenarios.

Of course there are several assessments forms whose objective is to provide a guide about how to execute this kind of evaluation. The problem with these forms is that you need to have some time and somebody has to answer it for you. In my professional life, several times I was faced with situations where I had very little time and no one to answer my questions. This item deals with this kind of situation trying to provide some sort of guideline to situations like that.

The first thing which anybody who wants to make an assessment of a large telecom infrastructure in a quick way and with a minimum help from the people involved has to do is to make an extensive research in the company itself. That means you have to get there knowing everything publicly available about the company:

- Companies which integrate the conglomerate
- What each company does
- Number and location of its sites
- Its modus operant (Which may or may not be similar to organizations in the same niche)
- Number of employees
- Services available through web and through 0800
- Visit a site and see what the organization has there.

Auditing reports which are usually available to publicly traded companies are also valuable sources of information.

With all this information mapped is time to try to evaluate the big numbers of the operation, things such as:

- Total expenditures with telecom (divided by type or provider)
- Main contracts (and with who)
- Number of people in the telecom team

Note that both groups of information are accessible bypassing the telecom manager or the IT manager. Sometimes this is crucial.

Here is where experience counts. If you have enough market experience, at this point you will be able to have a general view about how they operate and if their expenditures are within of what would be typical.

And finally a phase which you may ask questions directly to the people involved. Here I tend to prefer a more subtle approach than using a form with questions (If you have a form very often the person just ask you to live it there or e-mail it to him/her and he/she will return it to you filled). I usually try to stablish a conversation with the people involved where I ask more general questions starting for the ones which I already know the answers (how many sites do you have? for example). Here my objective is less to get the right answers but to evaluate how fast and how confident the people provide it. For instance if I ask to a telecom manager what is the total monthly expenditure with telecom, it is more important to observe if the telecom manager is used to provide this information and how quick and detailed this information is provided. Either if he/she has to consult papers and notes or know things by heart both behaviors are important cues.

I usually ask general questions such as "Please, tell me about your telecom infrastructure". Here once again what matter is not so much the answer itself but how the answer is given. Evaluating in which subjects he/She expends more time or give more emphasis and details gives you a good understanding how the infrastructure is viewed and managed. For instance if he/She starts giving a pure technical description with very few references to contracts and cost you can easily perceive that there is some problem (The opposite is also true). Other aspect to be observed is how often links between technology, costs and the business issues are made during the answers.

Other aspect to be observed is about the great numbers. If the answers given by the people interviewed doesn't match the big numbers provided by other areas of the organization that is definitely a red flag, be sure you get to the conversation phase knowing at least some of these big numbers. For instance if the person tells you he/She expends USD 1.000.000 monthly and the bill payment told you the expenditure is around USD 2.000.000 there is a problem.

Once again we have to remember we are not really talking about a complete assessment but a quick assessment. Very often you have only one visit for a couple of hours to figure out a general view and spot or guess the main issues. Not an easy task.

Other important aspect in these conversations is to be attentive to anything which sounds a bit out of normal. During these talks be aware of anything which for any reason doesn't sounds right. I know it sounds a bit common sense but considering that sometimes important decisions have to be taken based in so little information this "inner gut" felling matters. After working in the field for long time you develop a kind of six sense instinct which usually guides you through these processes. It works as long you learn to believe in yourself.

14 How to Deal with External Consultants

Very often the people in charge of the IT infrastructure have to deal with external consultants; the objective of this item is to discuss this relationship exploring the different perspectives in this interaction.

Consultants as auditors are not usually loved by the IT teams and the telecom area is not an exception. This dislike for external consultants usually comes from four typical sources:

1. It is natural human behavior to avoid the possibility of being exposed as running non optimal operations or to be put in a bad light. A consultant is always an outsider looking for something which is not good or could be better.
2. It is also a natural behavior to resist to initiatives taken without participation (what usually is the case when bringing in an external consultant).
3. External consultants sometimes behave badly, treat people with lack of respect and think about themselves too high (Who never had a bad experience with a young just graduated MBA). This is a very threating behavior to the internal folks considering that these people have the ears of the high management).
4. External consultants sometimes have their own agenda and try to justify their presence by looking for problems where they may not exist.

Therefore is quite common that there is a lack of cooperation between the external consultants and the internal team. Often there is a reluctance to be open and frank with consultants particularly if their entry into the organization is via some higher management function (What usually is the case).

In summary the problems in this relationship is often because people feel threatened by what results this process might show and whomever brought the consultants into the organization does not get the lower folks onboard with what he/She is trying to achieve.

Having worked as Telecom manager and as a consultant I am always stunned by how antagonistic these interactions frequently are. It is my view that they don't need or should be in that way.

I am not saying that being audited or having your area scrutinized by an external consultant is going to be ever a comfortable experience. It doesn't supposed to be. However it can be a very productive and growing experience depending on your approach of the situation.

First think about the process as a consummated fact, the consultants are there and there is nothing you can do about that. You have to make the most of it. Don't expend time or effort questioning the reason/wisdom for the auditing or consulting, it only makes you look bad.

Second try to think about the presence of the consultants as a signal of prestige of the area (It may not be true regards you personally but as a mental attitude it doesn't matter). If you think about that, someone in the organization decided to pay to somebody to look into the problems of the area in order to try to improve it. There is a said "Everybody in the future is going to have at least five minutes of fame", the consulting engagement is your time in the spotlight. Of course, be on the spotlight may not be a good thing but can be.

Third keep in mind that a consultant is somebody who tends to be facts orientated, is not there to be your friend, doesn't play organization politics and wants to find ways to make things better (At least the good ones). Try to be honest and neutral. The consultants usually are trying to understand the current scenario in a very short time span and compare it with some reference identifying the gaps. A smart consultant knows the importance of having your help.

Finally think about this as an opportunity to achieve the following:

1) If you are not sure if you have a good handle on what are the area problems try to achieve this view from the consultant, aligning yourself with the initiatives to solve them. Use the consulting engagement as an opportunity to learn and grow.

2) If you already have the points where you know the area can be improve don't be afraid to share your views with the consultants, they can be your voice in the higher management. This is your chance to get the help you were asking for (Never look the consultant as somebody with whom you are in some sort of competition).

Both topics depict common scenarios and in both cases you can come out of the process better than before. From the perspective of the consultant both approaches are also good, after all everybody likes to fell important (item 1) and it is a lot easier if the internal folks already have a "shopping list" of needs (item 2).

The approach should be similar to an audit, although most people hate it and are scared of the outcomes, a better way to look at that is it forces a company to deal with issues once exposed in an audit and usually provides the right visibility at the right level to make fixing findings a priority.

A good consultant usually tries to bring the tension down making sure that everyone understands their role and particularly what the objective of the engagement is. These approaches help to allay people's fears that he/She will show the current operation in a bad light.

A good consulting engagement is not the one where lots of problems are identified but one where problems are identified and the whole team goes onboard to fix them and adjust the processes to avoid them to happen again. A good consultant is not the one who just finds the problems and points how to fix them, but the one who knows how to empower the organization team to become able to perform permanently better. A effective consultant should know that he/She is transitory but the results of his/her work must last.

15 Why Use External Consultants

An interesting point worth discussing is the benefits of using external consultants. Although most telecom managers usually rebuffs the idea of bringing in external consultants we believe that external consultancy is a very beneficial process which aggregates a lot of value if properly conducted.

Here it is important to specify what we mean by "external consultancy". In Our view Consulting means people whose objective is to analyze the current scenario and prepare a diagnosis of the current situation pointing gaps and proposing solutions. The definition is important because today a whole group of activities usually define themselves as "Consulting services" what in our view is a bit misleading. For instance we don't see people linked with bill auditing, hardware installation/configuration and tariff renegotiations as providing consulting services. In our view these are specialized functions which may or may not be outsourced.

Talking about consulting services in this narrowed definition we can list pros and cons of a company hiring an external consulting team to review the current telecom scenario:

Pros:

- Access to broad networks of resources and expertise.
- Access to skills and capabilities that the organization may not have internally.
- Improve diversity and reduce internal bias through an external perspective.

- Facilitate decision making processes and mitigate internal competition.
- Access knowledge from other organizations and industry best practices.
- Build capabilities (skills transfer) within the internal team (mentoring).
- Isolate internal management from sensitive internal decisions.

Cons:

- It may take some time for the consultants to understand the environment.
- The consultants have limited stake in the outcome of the recommendations.
- Using external resources can potentially undermine the self-esteem of existing staff.
- Hourly rates and travel costs can be expensive on the short-term.
- The consultant may not have the skills you are looking for - choose carefully.

In general these are the reasons pro and against an organization hiring an external consulting firm and these reasons are not specific to the telecom area. Of course, the balance between benefits and costs depends of the specific context of each organization, but in general if well conducted an engagement to map the current scenario and compare it with an ideal situation identifying the gaps and preparing a plan to bridge them is usually money well expend.

Detailing the items above to explain why external consultants can be a very useful tool we could organize the reasons as following:

1) The external consultants can focus in analyzing the structure as a whole, detached from the day-by-day activities and from the politics of the organization. Here it is important to emphasize the point that most what they are going to do wouldn't be beyond the internal team own capacity, if only they had the time. Of course there are the consulting companies which deploy analytical tools

which are a way beyond what is possible to be achieved internally but those are the exceptions (traffic analyzers and topologies simulators for instance).

2) A consulting engagement gives the telecom team a chance to get away from the pure operational focus where the pressing demands usually force them into. Think about that what you don't do to improve permanently things due the fact that you are always oppressed by the "keep the lights on" approach. A consulting engagement is an opportunity to compensate that and a chance to think beyond the purely operational perspective.

3) External consultants do this kind of job in several companies and they usually bring a wider view about how things are usually done and how they should be done. This is an important point and a consulting engagement is an opportunity to tap in this knowledge. Not to mention knowledge about costs and tariffs (benchmarking).

4) Other important aspect usually associated with external consultants is the fact that these professionals usually are well versed in presenting complex issues in a simplified way being very effective in "Selling" to the upper management. So it isn't only about the gaps they actually found but also about the gaps that the telecom team already knew and was having difficult to make the management to give a proper attention to.

5) External consultant usually deploy some sort of methodology to organize and structure a plan to bridge the gaps identified (And even follow the implementation of the recommendations), this knowledge is by itself very important given the fact that the problems identified and the recommendations to solve them are organized in a formal and structured way, what is very instrumental to internalize the findings and align the whole chain of command of the organization. In other words is not only about the ability to find the gaps and the solutions is also about being able to

organize these finding and recommendations in a way which all organization can articulate itself to act.

6) External consultants for being expensive usually have the ears of the high management. It may seem an obvious statement but we do believe that there is a strong correlation between money expent with a professional and the attention given to his opinions. The correlation between money expent and capability is taken for granted (which may not always be true). This fact gives the findings the attention from the high management which they may not get if they were proposed by the internal team.

In summary it is my view that a periodic evaluation of the scenario (intervals between three and five years would be ideal) is a very important thing and can add a lot of value to the organization. Of course to be effective this process has to be well planned, the information necessary must be available and the telecom team has to make some time available for the consultants.

16 Where to fit the Telecom Bill Processing area within the Organization

A controversial issue is to where to locate the telecom bill processing area (telecom invoice processing) in the organization organogram (Regardless if the auditing is done in-house or outsourced).

Very often we see the telecom invoice processing process being executed outside the telecom area. Usually some sort of bill processing unit within the IT or even for a non IT area (the general bill processing area). In our view it is a very grave mistake and in this topic we are going to discuss why we believe that the telecom invoice processing has to be linked with the telecom area.

An apparently innocent decision to put the telecom invoice processing process to be executed for an area different then the telecom area is the root cause of several problems associated with managing telecom costs. This issue is made more insidious because the problems usually are not perceived immediately, being noted only when the telecom cost pattern of the organization becomes detached from what would be desirable/ typical.

Among the problems which this kind of arrangement generates, the worst, is the creation of a mindset in the telecom area where the main goal is to keep the structure running, the issue of the costs takes a back seat. This phenomenon happens along the time and is hard to spot and correct.

In this kind of scenario (Where telecom bill processing is executed by other area different than the telecom) we very often find situations where the telecom manager don't know how much the organization pays for the telecom resources.

Here it is worth emphasizing that we are not saying that such processes cannot be executed by other area (or even outsourced), what we defend is that if these processes are executed by other area (general bill payment area, IT cost management unit or outsourced) it should happen in a way where telecom keeps the control of the bills received, keeps the power and the obligation to approve (or not) each bill before it is paid and above all keeps cost as a fundamental KPI to evaluate its performance.

When we have a scenario where we have different areas executing different parts of the telecom cost management process the deployment of Telecom management ERPs become critical given the fact that these tools are enablers of integration among the several parts of the process.

Initially we have to keep in mind two basic points:

1. Invoice processing is only a small part of a larger process called telecom cost management, very often people talk about bill processing as if it equals to cost management, that is not true.
2. Invoice processing is the input for a good telecom cost management, not an end in itself.

Invoice processing is a mean not an end in itself. Therefore, this kind of process has to be organized in a hierarchical way placing the invoice processing as a provider of information to the management of the costs, being consequently subordinated to it.

If we look the topics of the TEM discipline in ITIL we can see that "invoice Processing" is just a small part (Although an important one):

Sourcing & Procurement	Invoice Processing	Auditing	Optimization
Service Ordering	Change Control	Contract Management	Asset Inventory Management
Service Inventory Management	Policy & Governance	Help Desk Management	Mobile Device Management
Mobile Application Management	Risk Management		Reporting & Analysis

The reason for emphasizing this point is due the fact that the interconnection of the processes is a defining factor to where to locate it within the organization.

The consequence of the described fact (subordination of the invoice processing to the overall process of telecom cost management) is that telecom invoice processing has to be linked with the telecom area. The reasons for that are:

1. The telecom area is where within the organization people know what is contracted and where the resources are.
2. The telecom area is where in the organization people know why the resources were contracted
3. The telecom area is responsible for the generation of demands for the providers, and therefore knows what was installed, what was cancelled and what was changed (with the consequent impact in the invoices).
4. The telecom area is (Or supposed to be) in contact with the telecom Market and is better placed to have a view of what are typical values and what are possible strategies to reduce costs.

5. The telecom area is responsible for the implementation of the telecom projects. Therefore is where the economic impacts of the actions taken can be better identified and evaluated. The costs variations may be understood in the context of installation or des-installation of resources or in the change of transport strategies.

Going one level further in the analysis of the topics:

1) The telecom area is the area within the organization which has (Or suppose to have) the inventory of resources contracted and knows where each one is and to whom in the organization the resource serves. This is a fundamental understanding to allow the approval of any invoice. The collorary is that a telecom invoice can only be paid if approved by the telecom area.

2) The telecom area knows the reason why each resource was contracted and whom the resource serves

3) The telecom area identifies the new needs (People may ask for resources or the telecom area itself identifies de need) and interacts with the providers to request new resources, cancel or change the existing ones. Therefore all changes in the invoices are known or are provoked by the telecom area.

4) The telecom area is where there are people capable to follow the technological trends, to know the typical prices for services and possible strategies to reduce telecom costs (Where there are people capable to evaluate if something is cheap or expensive).

5) Maybe the most important reason of all is the fact that the telecom area is responsible for the implementation of the telecom rearrangements (In a large organization there are Always something going on) which usually imply in change in the telecom costs. The telecom area is the only place where the correlation of cause and effect between technical/contractual actions can be fully perceived and understood.

For all these reasons it is crucial to involve the telecom area in the management of the telecom costs.

17 Auditing Bills – an Unbalanced Importance

When it comes to managing telecom costs one thing which strikes me is how much the bill processing and bill auditing processes are emphasized. It gets to the point where the term TEM (Telecom expense management) gets almost the same meaning of processing and auditing telecom bills. It is not unusual to have companies which call themselves as TEM solution providers when in reality they just provide billing systems or bill auditing.

If we look into the ITIL processes associated with managing telecom costs we have the following:

- Source & procurement
- Service ordering
- Service Inventory management
- Mobile application management
- Invoice processing
- Change Control
- Policy & governance
- Risk management
- Auditing
- Contract management
- Helpdesk management
- Optimization & traffic management
- Asset inventory management
- Mobile device management
- Reporting & analysis

As we can see even considering each item with the same importance in the whole process the relative weight of bill processing and auditing would be small.

But when we look how these items are actually ranked in terms of effort spent by the telecom teams when managing the structure we clearly see that bill processing and auditing usually get a much higher importance than it supposed to get considering the benefits and savings they yield.

Let's think about "Optimization & traffic management", which in our view is one of the most important items. If you manage your traffic well, doing things such as carrying voice flows through existing data network or selecting the least cost routes you probably is going to get savings a way beyond what is achievable through checking the bills. What is the point of checking the right price of a call which doesn't supposed to exist in the first place?

If we consider the items "Service Inventory management" and "Asset Inventory management": Knowing well what you have, where the resources are and why they exist, usually is a much more critical thing than checking the bills. What is the point for paying the correct price for a link which the organization is not even using anymore?

If we think about "Change Control" how can we guarantee that the items charged are correct if you don't control changes properly? You may or may not have changed a bandwidth of a link how to be sure the telco is charging correctly if the "service ordering" and "change control" processes are not working properly? How to know if the price is right if you don't manage your contracts properly "Contract management"?

So, we have here two types of situations: 1) activities which tend to generate more savings than bill processing and auditing and 2) processes which are prerequisites to processing and auditing the bills. As we can see, almost all processes fit into one of these two categories.

Once we had demonstrated this fact that there are several things, more important than or prerequisites for, bill processing and bill auditing the question remains: Why Bill processing and auditing get so much emphasis when talking about TEM (Telecom expense management)?

There are some reasons for that: 1) First is the fact that there is an operational imperative, bills have to be paid or the services will be cut. This imperative brings bill processing to the top of the priorities even if it doesn't really aggregate that much in terms of guarantying low costs (even if you count penalties for late payment). 2) Secondly there is the natural uneasiness about paying for bills which were wrongly charged. 3) Thirdly there is the fact that auditing bills usually generates quick and tangible results.

Other factor which somehow influence this unbalance is the fact that processing bills is perceived as a very bureaucratic process (What it is) and therefore something which would be better placed somewhere else in the organization (Not in the telecom area) or outsourced. That fact by itself pushes the process to the hands of the bill auditing companies what adds it in the BPO (Business Process Outsourcing) packet. The fact that this service is usually part of the BDO packet keeps it on the radar of the companies providing auditing bills/BPO and consequently keeps the telecom manager hearing about how good and how helpful this service can be. (Just to make the case, you don't see the same speech about traffic management for instance).

In addition to these reasons (operational imperative, uneasiness for paying wrongly and quick results) there are other two groups of reasons which also play a role in making bill processing and auditing so high in the list.

1) The first group could be summarized in a phrase "the other things are more difficult to do". That means, most of the items in the list are much more difficult to implement than treating invoices and checking bills.

    For instance analyze traffic is hard to do and demands not only dedicated tools but much more specialized professionals. Organize internal processes is also hard, takes time and demands interactions with several groups within the organization. Organize ordering and provisioning processes demand coordination not only with other groups within the organization but also with external providers.

2) The second group is linked with what we could call "conflicts of interest". Lets analyze the typical arrangement between the organizations and the bill processing and auditing providers. Either they are paid based on results (Percentage of the errors found) or a fix value based on the average gains their service generate. That means if they act in order to reduce permanently the errors found in the bills they are likely to hurt themselves. Organizing inventory and contracts, interacting with the telco's billing teams, organizing the ordering and provisioning processes all are activities which overtime tend to reduce the charging errors, all are activities associates with TEM but all produce results which collide directly with the interests of the companies providing auditing. What incentive a bill auditing company would have to work with the telcos to reduce permanently the errors spotted in the bills? In the worst scenario (When they are paid as % of the errors) the monthly revenue would fall and even if they are paid a fix amount monthly, soon enough the company would start asking itself if the service is worth the cost, considering the low percentage of errors.

If we look these reasons in an isolated way they seem to make sense (If not morally justifiable) from each actor perspective. However, when we look the issue as a whole we clearly perceive that there is a distortion.

It is our view that the bill processing and bill auditing have to be the last processes to be executed. Only after you got a good inventory, organized the contracts and tariffs, analyzed the traffic, organized the ordering and provisioning processes you can start effectively auditing bills. You have to work in the causes of the errors and act in the direction to minimize the errors. Ideally, bringing the percentage of errors to below 1%. At this point you can start auditing bills after the payment and control them based on historical value (Auditing bill every six months for example).

It is interesting to note that when seeing advertises of management/ auditing tools, auditing services and BPO (Business Process Outsourcing) services we usually spot phrases such as "Telcos usually overcharge you in average between 5% and 7%" or things like that. That implies that these percentage are immutable and there is nothing you can do about that. That is not true.

18 Billing Errors – Why they don´t fix the Problem

As mentioned before when seeing advertises of management/auditing tools, auditing services and BPO (Business Process Outsourcing) we usually spot phrases such as "Telcos usually overcharge you", "Telcos overcharge between 5 and 7%" or similar things. Those statements imply that these errors and overcharges are facts of life and there is nothing you can do about that. Consequently you need an auditing service or a control software otherwise you will be condemned to keep losing money. Reality however is a bit more nuanced.

Even though telcos are known for overcharging, there are several initiatives which can bring the percentage of billing errors down and most of them are not associated with auditing bills:

- Unify and simplify the contracts;
- Adopt flat rates;
- Simplify and standardize the mobile charging plans;
- Control better the assets guaranteeing that they are linked to few contracts;
- Simplify the minimum usage rules;
- Identify where the telcos are making the mistakes and sit down with them to discuss how to solve them.

Bill auditing companies tend to focus in identifying the billing errors instead focusing in finding the root causes of such errors and acting to solve them. A naive observer may argue that the responsibility for fixing the causes of the billing errors belongs to the telcos or the organizations not to the bill auditors. This view maybe partially true but do not describe the whole story.

My understanding of the situation is that both the organization and the telcos have interest in avoiding billing errors, the bill auditors (middle man) do not. This fact generates a situation where who has the means to identify the causes of the errors and discuss such causes with the telcos, usually has no interest/insentive in doing so.

As already mentioned the auditing companies have a "conflicts of interest". Either they are paid based on results (Percentage of the errors found) or a fix value based on the average gains their service generate. That means if they act in order to reduce permanently the errors found in the bills they are likely to hurt themselves. You should be aware of this fundamental conflict and avoid it.

Organizing inventory and contracts, organizing the ordering and provisioning processes all are activities which overtime tend to reduce the charging errors, all are activities associates with TEM but all produce results which collide directly with the interests of the companies providing auditing.

The organizations have to be aware of this fact and tread the issue wisely. It is our understanding that the first thing is to audit the bills identify the percentage of mistakes usually found and the amount of money it represents. This is the basic information, necessary to evaluate the size of the problem and consequently the strategy to be adopted. If the percentage of errors exceeds 1% the second step is to identify why the mistakes are happening.

Usually charging mistakes happen due factors such as too many contracts with too many tariffs, negotiated tariffs not being applied by the telcos, calls wrongly classified, etc. You have to map the mistakes identifying the percentage they represent. Once you have identified the typical problems, separate them between two groups:

- The ones you can solve yourself
- The ones which demand the participation of the telco to be solved

The ones you can solve yourself encompass things such as consolidation of contracts, organization of the provisioning process and simplification/standardization of charging plans.

The ones which demand telco participation encompass things such as wrongly deployment of tariffs and wrong call classification.

Sit-down with the telco billing team and explain the findings and try to work out a plan to solve the main issues identified, defining deadlines and verifications. Audit the bills and check if the percentage of errors is going down. This whole process usually takes between two and four months. The target is to bring the percentage of errors down to below 1%.

Once you have brought the errors percentage to below 1% you can start auditing the bills every six months (the time frame can be different). Auditing every six months doesn't mean you are going to pay de bills without any verification you must compare the bills against the historical values and any deviation above 20% of the average must trigger an auditing process.

It is our understanding that bill auditing is an important service however you should be realistic about what can expected from it. Bill auditors have no incentive to solve the root causes of the mistakes reducing permanently the percentage of errors in the bills. If they are paid by the % of errors spotted the monthly revenue would fall if they do that, and even if they get a fix amount monthly, soon enough the company would start asking itself if the service is worth the cost, considering the low percentage of errors.

Reducing permanently the percentage of billing errors have several benefits, one which is particularly prize worth is the fact that if you don't need to audit the bills before payment (Because the average level of errors is low) you eliminate a process which generates a lot of stress because you don't have to get the bill audited before the invoice due day (this timeframe is usually between two e three weeks).

19 Considerations about Automatization of the Bill Processing

When talking about processing the bills very often we see people defending the complete "automatization" of the process as a kind of panacea, automatically downloading bills and automatically uploading it to bill control applications would be the final solution.

Of course if we could have a process where the bills were gotten automatically and automatically recalculated and when discrepancies are spotted or if the totals differ too much from what is the typical it is automatically signaled by the application that would be great.

However, this ideal scenario has some problems: The first one is the obvious fact that a large organization usually has hundreds of telecom bills, usually; many of them are delivered in paper.

Secondly, even if all bills could be delivered in magnetic format, surely not all telcos would be able to make them available to automatic downloading in their websites. Some surely would be delivered by mail stored in CDs.

Thirdly even if we could have a scenario where all providers could make the bills available for automatic downloading, we still would have a good effort to adjust the procedures for each provider (which may vary widely).

Fourthly, Even we manage to solve all previous three issues we still would need some sort of synchronization process to make sure that all bills were made available and captured within the right time frame (Including recurrence in case they were not available) and to check why they are not available (How to deal with non-conformity).

And finally there is the issue of the changes in the contract inventory, in large organizations contracts are canceled and new ones stablished all the time, that means all the problems previously mentioned will be revisited periodically.

Even if you succeed in navigating through all these five types of issues it is necessary to evaluate the tradeoff between the additional complexity and the gains in the quality of the control.

Typically, in order to make the process free from human intervention (automatized) you may demand a much more specialized professional to take care of it. The situation is pretty much like automatizing production through the deployment of robots. It is only feasible when you have a big enough demand and a reasonable standardized process. As it happens when deploying robots you usually dispense low training personnel but increase the importance of high skilled professionals to take care of the robots.

Given these facts it is easy to perceive that this scenario would only be feasible if we had very few and standardized bills (what could be a reality if the organization had its business concentrated into few contracts and providers), and a big enough demand.

However, this scenario is rare and therefore is hard to advocate an operational model which although looking ideal in theory is so far apart from what is typically found in real life.

But even if we manage to have a scenario where we effectively could get all bills and process them automatically, we have to keep in mind that the human eye is the ultimate cost control instrument and any process which completely takes away this instrument tends to reduce the effectiveness of the control. The idea that the bill processing can run in automatic pilot mode may not be a good one.

The tools should enhance the human capacity of control (highlighting points of attention for example) not try to replace it. This is an important point because touches a sensible aspect which is what exactly people understand by "controlling" telecom costs.

There is a verbal trend in the marketplace which limits the meaning of the definition telecom expense management to managing bills and billing. This trend is driven by the bill auditors and tools providers who tend to emphasize these particular parts of the process to the point that these parts look like they were the whole thing.

Telecom expense management encompasses much more than just managing telecom bills and contracts. Telecom Expense Management (TEM) has almost the same meaning of Telecom Cost Management (TCM). Managing telecom costs goes a way beyond just managing the bills. The auditing of the bills maybe completely automatable but the understanding why specific costs went up or down, How the traffic behaved and the development of the overall costs and usage needs to be followed by a human.

Let's imagine an example where we have a low skill professional in charge of inputting bills into a telecom management ERP. The telecom manager may see this procedure as a complete mechanical work, which could be completely automatized. However when we look closer we see that this person may do other things which are more difficult to automatize such as if the bill didn't come look for the reason, if a new bill appears try to understand why (maybe a new resource was contracted and nobody informed it to the system). The person may even be able to see or know the typical values and spot big variations from what is typical (Increases or decreases).

The point is that depending of the context, this person conjugated with an external bill auditor may be as efficient as a complete automatized operation, maybe costing less, and above all be easier to replace.

On the top of that there is other subsidiary aspect of this issue which is the fact that the root cause of the billing errors should be identified and addressed, bringing the percentage of errors down permanently. Creating a process which alleviates the weight of verifying the billing errors may not be stimulating to make people search and solve the root causes of those errors.

20 Invoice Processing – How to Manage the bills

In this topic we are going to discuss the invoice processing. This is an important topic because many telecom managers see it as a process which doesn't belong to the telecom area and should be executed by another areas of the organization or outsourced.

As we discussed in the chapter about the need to have telecom technical and financial control unified we believe that telecom has to be involved with managing telecom bills even when not executing it directly. It is our understanding that finance should do the AR and AP function, however all the telecom bills have to be approved for payment by someone in the telecom team.

Here we are going to explain how we believe these processes should be executed and doing so we hope to make clear to the reader the difficulties which may exist. First we have to understand the two basic steps of processing telecom bills:

1) Getting the bills
2) Verifying if the bills are correct

1) Getting the bills

It may seem a straight forward process, however we have to remember that in large organizations we may have hundreds of bills which sometimes come from several countries and are priced in several currencies. In addition of that they may take several formats:

- Paper – The bills may be delivered by mail having thousands of pages detailing the calls (CDRs)
- Electronic media – The bill may be delivered in electronic format files recorded in CDs and the CDs delivered by mail
- Electronic media – The bill may be made available in the providers website to be downloaded by the client.

Summing to the difficulties telecom bills may be delivered in different locations and in different levels of grouping (one by resource or one encompassing several resources). It is worth noticing that some bills are sent by the providers to the organization but others are just made available and therefore it is up to the organization to go to the provider website and download them.

Even when the bill is forwarded by the provider, the provider may send it to the address where the resource is installed and not to where the bill is processed and therefore it is up to the organization to make sure that the bill gets to the right location within the organization.

That said becomes clear that in large organizations just making sure that all bills were received demands some effort and above all some knowledge about which bills supposed to be paid monthly and when and how they are gotten. It is absolutely basic to be able to notice if any bill goes missing.

In addition of that there is some effort/knowledge involved in downloading the bills made available in the providers websites. The professionals in charge have to know how to navigate each tool, depending of the case it can be a challenge in itself.

Make the process of getting the bills easier is one of the reasons why we should try to unify the contracts as most as possible, groping the resources in fewer bills as possible.

For instance, if we have a scenario where you have an organization with only one provider and two hundred sites but each site receives his own bill, we would have an situation where even to put all these bills together would be a big logistic effort. In this context, imagine how much trouble can be saved just consolidating these bills in only one bill and forwarding it to one address.

Here we have to remember that resources (with correspondent bills) are contracted/ canceled and prices are updated all the time and the people in charge of getting the bills have to be able to keep track of whatever changes. It is crucial to be able to know if a new bill has to be gotten or a previously existing one don´t have to be looked after anymore. That knowledge demands a close interaction with the telecom team which is the group within the organization in charge of doing these changes.

2) Verify if the bills are correct

Once we get the bills we have to verify if they are correct. This verification can be done with different levels of deepness.

In a more high level we need to be sure that we are paying only for resources belonging to the organization and making sure that everything which was included, changed or canceled during the previous charging cycle is reflected in the bills. Although it may seems a bit obvious, the first and basic verification (very often not executed) is to check if the bill belongs to the organization, and the resource is actually being used. In large operations with several sites (sometimes in different countries) it is not uncommon to receive bills referring to resources that don't belong to the organization or are not in use anymore. Subsequently we can go deeper and verify things such as:

- If the value charged is compatible with the historical value.
- If there are undue charges for installation or subscription
- If there are penalties for late payment unduly charged
- If there are charges for not achieving the minimum committed volumes (voice)
- If the taxes (or taxes exemptions) were applied correctly

Going one step deeper we can verify if there are credits due reimbursements for operational problems or due billing errors in the previous bills. Going even deeper we can verify if the costs of the voice calls and mobile services were properly calculated and reflect what is defined in the contracts (It is what is usually known as bill auditing).

Note that all previous items demand knowledge of the contracts and intimacy with what the telecom team did in the previous month, including maintenance problems, canceling of resources, changes and installation of new resources. Not to mention negotiations and billing disputes.

How to structure the bills verification

Here we have several possible strategies to execute the verification, varying from verifying only the historical values (Simpler one), passing through checking the basic points listed to the complete recalculation of the voice bills. Note that we may adopt a mix of strategies. Example: Checking the historical values of the bills of smaller values, checking the parameters such as undue charges, penalties and reimbursements for the big voice and data bills and completely recalculating the calls only for the big fix and mobile voice bills.

Here it is interesting to note that we usually have some typical types of procedures regards recalculating the voice bills (What is usually known as auditing): 1) Do it for all bills every month 2) Do the auditing only when the value charged differ a given percentage from what is typical or 3) Do a complete recalculation of all bills every predefined period (Each six month for example).

Some organizations don't do the complete recalculation of all calls every month, doing that only for the ones in which big discrepancies between what was identified by the billing system or historical values and what was actually charged were spotted.

The telecom team has to evaluate the balance between the effort vs benefits of the verification, and also take action to fix the root causes of the billing mistakes, that is one additional reason to keep this process close to the telecom team. The telecom team is not only the people able to understand the process as a whole but also the ones in charge of the dealings with the providers both in technical and financial terms.

Disputing the discrepancies

Regards the process of disputing the values charged we have three typical approaches: 1) pay the bill as charged and require the reimbursement of the discrepancy in the next month, 2) pay only the value considered by the organization as correct and let the discrepancy to be discussed later or 3) don't pay the bill at all until the correct value is identified. We believe that the second option is the fairest. Again the participation of the telecom team in the dispute process is crucial due several reasons:

1) They are the ones who negotiate the contracts and Know the organization needs and are able to compare tariffs
2) They are the ones who control the technical problems and know the impact which a bad quality of service can have (Penalties in values of data links are usually associated with total downtime during the month).
3) They are the ones who maneuver the contracts and can measure the effort to migrate from a provider to another
4) They are the ones aware of the initiatives in curse within the organization's telecom infrastructure and positioned to evaluate the impacts of the measures taken.

A typical scenario is having a specialized team (within the organization or outsourced) collecting and verifying all the invoices. Then they pass a package to the telecom team to be approved (A required step). If there were erroneous invoices this team would highlight what was wrong and the telecom team would be responsible for dealing with the telcos.

As we can see it is typical having two different teams (telecom and bill processing) the bill processing team usually working subordinated to the telecom team (often having the telecom manager as the coordinator of both groups). The key point however is how you make these two groups of people work together properly. The bill processing team depends heavily on the telecom team to know everything which may change in the bills including but not limited to contracts management, inventory control and tariff changes. On the other hand the telecom team depends on the bill processing team regards everything associated with bills disputes and costs verifications.

This is a symbiotic relationship and the challenge is to make these two groups work together well (regardless if they belong or not to the same area within the organization or are outsourced). Of course if the telecom manager is the leader of both groups things tend to go smother. In the same way if the groups belong to the same organization (Are not outsourced) and work side by side they tend to be more integrated. If they deploy the same telecom ERP where they share information such as inventory of resources, contracts and price lists and service providers interactions the integration also improves.

In summary we could guess the level of integration using the following spreadsheet:

Item	Yes	No
Telecom manager is the leader of the two groups		
The bill verification and telecom team belong to the same organization		
The two teams work side by side (Physical proximity)		
There are formal milestones defining the interaction between the two groups		
There are a Telecom ERP integrating the two groups regards inventory of resources		
There are a Telecom ERP integrating the two groups regards contracts		
There are a Telecom ERP integrating the two groups regards bills		
There are a Telecom ERP integrating the two groups regards service providers interations		

The more "YES" we have more smooth the two teams tend to work.

21 Telecom Management in an Evolving Scenario

The objective of this item is to discuss the relative importance of telecom within the IT area of large organizations along the time. How it is evolving and to where it is going. Working in this field for almost thirty years now I saw lots of changes and taking a step back I can perceive some trends which maybe give us clues to where we are heading.

Looking back we can see big changes like when the analog PBXs were replaced by digital ones, when the use of e-mails and internet become widespread, when the integration of voice and data become common place or when the mobile services started. All these technological changes impacted enormously how telecom was used and managed by large organizations.

But the idea is not so much to discuss the technological changes themselves but how those changes impacted and continues to impact the way telecom is charged and managed. It is interesting because somehow the management and charging lingers behind the technological changes. How the organizations are evolving to adapt to this new context, not only in technical terms but also in management.

First it is important to make clear how we see the future. It is important because it will allow us to speculate about how the telecom management will evolve. It is our view that the key technology (We may call it disruptive) is cloud based services including SAAS (Software as a Service) and infrastructure. The consequence of this is the fact that Internet services are becoming critical from an internal access perspective (Although not for mobile and costumers users).

Following this trend we have an increase in demand for security services (how to be safe in the public place like the internet) and somehow a decrease in the demand for traditional infrastructure (including but not limited to telecom).

We see a scenario for the next five to ten years where the organizations will be struggling to adapt their particular IT needs to a hybrid mix of technologies which will spread between two extremes (internal and cloud) and the telecom infrastructure will follow this trend with software defined WANs. The more cloud based the organization is the more software defined the WAN will be.

We will have scenarios where you have accelerators that also make routing decisions based on policies rather than typical QoS and routing schemes, Internet connections will be understood exactly as internal connections, it will have implications on gateways back into internal datacenters.

Following this trend voice will become even more commoditized, and the TDM connectivity will lose more and more ground being replaced by SIP technology. It's already happening, but SIP vendors have being lacking in some services, but over time they tend to catchup.

Although SIP is already common place, we will start seeing more and more services such cloud based PBXs, and this trend, no doubt, will be very disruptive to the corporate market landscape for voice services and of course for PBXs vendors. Cloud PBX setup is likely just SIP services, maybe large call centers operations will take some time to adapt but administrative traffic of large corporations surely not.

Services such as skype for business will become big cloud PBXs, it has the same issue with call centers but as businesses move to using applications such as Office 365 Microsoft the services such skype for business will integrate naturally.

Think about applications which are full PBX capable environments and can be connected directly to, for example, Cisco voice gateways exactly as Cisco Comms Manager. This is the kind of environment which we are going to see more and more often.

Let's imagine scenarios, where you will have Instant Message and voice and active directory integrated. In the future onboarding and off boarding will become more and more easy as will the general provisioning cycle. In a near and medium future all this will run in the cloud (Ex: Microsoft cloud) and once you moved to applications such Office 365 then all which you will need is a client on your smartphones to have only one phone.

Assuming a scenario where most organizations would be using an office packet, Instant messages and voice all in the cloud. In this scenario would be relatively easy to connect these to a wholesale SIP provider in various regions. Then we would have the opportunity to really change the telecom cost structure.

In a sense our view of the future telecom world has an analogy with the real world, where in the future we will have a scenario where the organizations will do more and more things through others, relaying more and more in the network of providers and transporters to minimize their own cost. Where the commerce will be freer and the services and products will be made where is cheaper, minimizing costs and improving overall productivity. All of this gotten through the increase in complexity and decrease in self-reliance. In a nutshell the system will be more efficient though more complex and more fragile.

Of course all these views assume that we will have a reliable and cheap internet and this network will be able to handle all our needs without major's hindrances. As in the real world all sorts of problems may act to hamper this scenario (Ex: WARs, security treats or lack of network neutrality).

Other associated aspect of this view is that an old concept may come to age. "Cyber Carrier". Carriers will relay more and more in hosting services then in actually providing the connectivity itself, they will be the "cloud" through were everything flows and where the services are hosted.

Now we described how we believe the corporate telecom environment will be in the future we have to think how the way we manage it will change. One thing is certain, this new scenario pushes the management away from what is typical today.

If we think about the concept of cloud, somethings are noticeable, let's use AWS (Amazon Web Services) as an example, you have to setup your own network and security, within AWS. It is all about virtual appliances. This kind of process does not decrease the telecom manager workload; in fact it increases the complexity of the tasks. Tasks and processes today distributed across several providers (PBXs and router providers) will have to be executed by the home team (or contractors). Things which sometimes were left with the default configuration will have to be carefully planned and configured.

As already mentioned security will become even a bigger issue and everything will need to be encrypted, this is a security thing but execution probably will fall to the Telecom team. Cyber carriers will not be doing the encryption, the organizations will. SD WAN fits in perfectly with that concept. The need for more security policies will make the telecom area more complex to operate.

Today, several organizations already have telecom running edge and DMZ firewalls and to them this transitioning will be easer given that they are already familiar with the nature of part of the work of the future telecom area.

That said, today we already have in many organizations scenarios where the execution of security policies are not exclusive responsibility of the security team, as we mention the trend will be to transfer more and more security tasks to telecom. In addition of that the cloud infrastructure space also complicates life for telecoms teams as it adds complexity in areas not directly related to speeds and feeds as in physical connections. Of course those things will remain necessary however disputing space with the new scenario.

It is our understanding that in the future, less than 10% of what the telecom team does today will remain necessary, although lots of new things will become more and more important and will need to be done. The work will be more and more linked with configuration and security. In the end of the day it will add complexity.

We see a scenario where you may be able to operate "traditional" telecoms with lower tech skilled professionals, however when things break you will need professionals with more tech skills than is currently the case. In a way it will be pretty much like automatizing processes using robots. The whole things will run smoother and requiring less people, but if any problem occurs the people needed to fix things will need to be more qualified.

If we think about how the future role of the telecom manager will be, we see a scenario where configuration and security management increases in importance and difficulty and at the same time the whole process of provisioning, negotiating and controlling costs will become simpler (Commoditized). The relationship with the providers will become more indirect given the fact that it will become possible to contract services through the internet and even make quotations remotely as if buying commodities.

In this new context managing bills and providers will become simpler and managing the configuration of services in the cloud and managing internet accesses will tend to get an upper hand over managing, PBXs, fix lines, data circuits. Of course not all companies will be completely in the cloud and a whole range of gradation will coexist.

One of the key points which it is noticeable is the fact that the relative importance of telecom tends to continuous (Becoming even more critical) but now very much mixed with all sorts of other services. Voice services costs are going down (Even if Mobile costs are going up) and data services going up (As a whole).

22 Closing Words

I hope that this book proves useful as a reference guide. I tried to consolidate a large set of information in a coherent way in a relatively small space. In an endeavor like this, it is almost impossible to include everything which you initially intended; however, I believe I achieved my goal: to provide a useful tool through which IT and telecom managers can improve the efficiency of the management of their infrastructure.

Many of the topics discussed demand some previous understanding, and many of the opinions expressed may be somewhat controversial in their interpretation by the author. In addition, I am very aware that organizational realities may drive a politically based decision, which takes away the Cartesian line of thought expressed in this book.

I took the perspective of telecom managers; this book was written mostly for them. I know that many of the topics covered in the book may be viewed from a different perspective, depending on which hat you are wearing. Hardware vendors and telco representatives may also benefit from this book by improving their understanding of the challenges faced by their clients.

23 Acknowledgments

Many people contributed to this book. There no way I can mention all people who have taught, inspired and helped me – be it managers, colleagues and clients. Thanks to you all.

I need to specially thanks my longtime friend Ben Naude for reviewing most chapters, always contributing with good insights.

Finally yet importantly I thank my daughter Debora for being a source of inspiration and joy which allowed me to keep the disposition to develop this work.

24 Bibliography

Bayer, Michael. CTI Solutions and Systems: How to Put Computer Telephony Integration to Work. McGraw-Hill, 1997.

Michael Brosnan, John Messina. How to Audit Your Bills, Reduce Expenses, and Negotiate Favorable Rates. CMP Books, 1999.

Network Analysis Corporation. ARPANET: Design, Operation, Management, and Performance. New York: 1973.

Software ARIETE®- WANOPT®, 2010.

Software TRMS® Telecommunications Resources Management System- WANOPT®, 2014.

S.C. Strother, S. C. Telecom Cost Management. Arthech House, 2002.

Wide Area Network Methodology® - WANOPT®, 2010.

Printed in the United States
By Bookmasters